HOW TO FIND A JOB AND KEEP IT

IN MEMORY OF ANNETTE

AND FOR TRACEY, HOPE, JAMES,
SARAH, JOSEPH, HANNAH & EMMA,
WHO ARE ON MY HAPPINESS ROAD

HOW TO FIND A JOB AND KEEP IT

SIMON BOYLE

EBURY
PRESS

3 5 7 9 10 8 6 4 2

Ebury Press, an imprint of Ebury Publishing,
20 Vauxhall Bridge Road,
London SW1V 2SA

Ebury Publishing is part of the Penguin Random
House group of companies
whose addresses can be found at global.
penguinrandomhouse.com

Penguin
Random House
UK

First published by Ebury Press in 2016

www.penguin.co.uk

A CIP catalogue record for this book is available
from the British Library

ISBN 9781785035005

Original design by Alison O'Toole
Typesetting by Jerry Goldie

Printed and bound in India by Replika Press Pvt. Ltd.

Penguin Random House is committed to a sustainable
future for our business, our readers and our planet.
This book is made from Forest Stewardship Council®
certified paper.

MIX
Paper from
responsible sources
FSC® C018179

CONTENTS

AUTHOR'S NOTE

THEY SAY LIFE IS VITAL. WELL, ACTUALLY,
IT'S THE JOURNEY THAT SHOULD BE VITAL —
FILLED WITH GOOD ENERGY, SKILLS
AND, OF COURSE, LOVE.

THIS BOOK IS DEDICATED TO ANYONE
WHO HAS BEEN LOST, BUT WHO WANTS
TO SEEK A NEW PATH TO A BETTER,
HAPPIER AND MORE PROSPEROUS FUTURE.

PLEASE READ IT AND USE IT OR GIVE
IT TO SOMEONE ELSE WHO NEEDS IT.

THE VITAL JOURNEY

CAPACITY TO SURVIVE

POWER TO LIVE, GROW AND DEVELOP

BE A FORCE TO CONTRIBUTE TO SOCIETY

GAIN MEANINGFUL EMPLOYMENT

STRIVE FOR A PURPOSEFUL EXISTENCE

ALWAYS REMEMBER, HAPPINESS IS NOT AT THE
END OF THE ROAD, IT IS THE ROAD

INTRODUCTION AND PURPOSE

Going out to work is hard. It means taking responsibility for yourself and paying your way in society. It can be difficult and complicated. You're accountable – success or failure is down to you. But once you're on top of it, nothing else makes you feel prouder, stand taller or gives as much satisfaction.

This book is intended to be a wake-up call, a poke in the ribs, a helping hand and a source of support as you move forward in your working and personal life. You're leaving behind a a life where you need handouts, suffer low self-esteem and lack confidence – in order to create a life where you can do something you're proud of, make a good contribution to society and leave a positive legacy.

If you're on the edge of that important decision – should I work or not? – then this book is for you. It might not feel that way at first, but read on. I'm not trying to tell you what to do, but just share some of my experience. It's a guide, a simple arm around your shoulder from a mentor who tells it like it is.

Work is a massive part of life – usually you see your colleagues more than you see your family and friends. So it's important to choose your occupation carefully. It needs to be something you'll enjoy. But don't forget: a job doesn't have to be for life. It can be a stepping stone towards your desired destination. Look at a job as a rung on a ladder, with each new position taking you a step higher towards your goal. Sometimes it can be helpful to take your time on one particular rung. At other times it might be better to go down and start again in order to build up your confidence. Or you might need to step off for a while before returning to make new progress.

One thing is for sure: unless you're prepared to get on the ladder, it's difficult to reach new heights.

The obvious thing to remember is that a job enables you to earn your own money. It means you can stand up proudly and pay your way in society. You can afford to enjoy life a bit more and even start saving for the future. There's a whole new world waiting to open up for you.

WHY IS IT IMPORTANT TO WORK?

(AND STAY IN WORK)

There are many reasons why work is important. Financial stability means that you can live life to the full on your own terms; it prevents you being a drain on society, and enables you to take important steps towards a secure future. But most importantly, work enables you to live life with a purpose.

It's not always as simple as it sounds, and there can be many reasons why people struggle to hold down a job. But it seems to me that the failure to understand why work is important is the biggest hurdle preventing people getting, or staying in, a job.

LIVING LIFE WITH A PURPOSE

The idea of living life with a purpose is a basic one, but so many people seem to miss the point entirely. I believe it is vital to get up in the morning knowing you have something to do that will contribute to your own well-being, the success of your employer and the functioning of society as a whole.

No matter what it is, getting up full of energy and taking part in something is better – much better – than doing nothing. Nothing has ever been achieved by doing nothing. I have heard many successful people recall the day their life changed, and their stories usually start with them getting out of bed and doing something simple: 'I wonder if I hadn't got out of bed that day . . .'

DON'T BE A DRAIN ON SOCIETY

This is a tough one. We all need support, and at times that possibly has to come from the state, meaning the government and the people of this country. But staying on benefits and becoming a statistic is no good for anyone long term. If in the short term it helps you refocus and provides a financial safety net while you work out how to get yourself together and take the next step, that's fine – that's why we have a welfare system. If you end up in long-term unemployment for health reasons or you have some massive barriers to employment – for example, you have had a period of homelessness and lack confidence and self-esteem – again people will understand. But, and it's a big but, if deep down you know that with the right support you might be able to stand on your own two feet and work for a living, then you ought to try.

There are always going to be difficult times for everyone. We all have the government, banks and energy companies trying to get more money from us. Life just costs more and more, doesn't it? Benefits are being cut and living off the state is harder than ever. So if you are able to work and help yourself, it is hardly fair that others have to foot the bill for your benefits when they are also struggling to live comfortably.

LIVING LIFE TO THE FULL

Life is all about choices. For me, there's a conscious decision to be made. Do you want to enjoy life or simply go through the motions? I know it's not always as simple as that – personal problems, health issues and financial challenges can stop many people from thinking positively and living life to the full. But if you at least make the decision that you'd like to try, that's a start!

We tend to judge our success in life by how much disposable income we have to play with. But, really, life is about happiness and love, both of which shouldn't cost a penny. Of course, we have to pay the bills, put away a little in savings and hopefully have some left over. But add happiness and love into the mix and suddenly every day life can be lived to the full.

WORKING TOWARDS A SECURE FUTURE

Once you have decided to contribute and live your life to the full, then you can start to think about the future. I have always struggled to save for the proverbial 'rainy day', but it's something we all need to think about. One day something might happen in your life that means you need some extra money. It's therefore worth working towards saving a little extra in a reserve account for when that day comes.

WHY IS IT SO HARD TO STAY IN WORK?

Getting into work is one thing, staying in work is another altogether. Here's the answer:

FIND SOMETHING YOU LOVE TO DO.

THAT'S IT. SIMPLE, EH?

If you watch most people on their way to work, they seem to be having a personal nightmare each morning. They're trapped in jobs that they hate. In order to make employment work for you and enable you to stick at it, you have to enjoy what you do and look forward to it. Value your work colleagues. Take satisfaction in your job each and every day, otherwise it becomes a bore, a headache and, worse, something that you could lose if you don't do well at it. Learn something you enjoy, get stuck in and do it well, and the benefits can be fantastic.

Of course, it's easy for me to say. I was born to be a chef and I have always cooked. So many people simply don't know what they want to do, let alone have a personal vocation.

It's worth trying a few things to see what gets you excited. Maybe offer yourself out for a few work trials. Do it for free, just to get a feel for different occupations. You'll be amazed to see that lots of employers want to help people out, so explain what you're trying to do and you might get the chance to find something that really appeals to you.

HOW TO USE THIS BOOK

As you have committed to getting into work, this book is designed to give you a thoughtful and purposeful shove in the right direction. It is based on my own experience and opinions. These won't be shared by absolutely everyone, but whose are?

I want this book to help you to act differently, to provoke you to get up and change your life. You can do it! I also want to support you as you make these changes. As you work your way through the book, I hope that the importance of finding and sustaining good, meaningful employment will become clear.

Yes, you need to be brave, but if you're reading this you've already taken the first step. Don't procrastinate and set the task aside. Don't put the book down and lose focus. Don't think, 'I'll change tomorrow.' Read it now and put my suggestions into practice. You will be better off for it.

By the way, we all need help from time to time. Even the most successful people – in fact, especially the most successful people – have mentors, supporters and people who look out for them. When I say successful, I mean influential people who are financially and/or academically successful. They will have people who support, guide and advise them. They will not be going it alone. They will be eager to learn despite their position and experience, but will be thoroughly supported in order to achieve their goals.

So don't think you're on your own. This book will open up for you the idea of seeking the right counsel. See the book and my writing as a kind of personal mentor. You don't need to put everything into action straight away but as and when it's appropriate for you. But do read it and make notes. I've left space for you to do so. Keep it handy. Share your thoughts with others, and I promise it will help you to make significant changes for the better in your life.

Personal Projection Plan

On the following page is an activity that can help you understand yourself better. Like all activities, the more honest you are, the more you will learn about yourself. Try to write at least a full sentence for each section if you can.

PERSONAL PROJECTION PLAN

DESCRIBE YOUR CURRENT SITUATION AND HOW YOU FEEL ABOUT IT.

....

....

WHAT IS HOLDING YOU BACK?

....

WHY DO YOU THINK YOU NEED HELP?

....

WHAT ARE YOUR HOBBIES?

....

WHAT ARE YOU INTERESTED IN?

....

WHAT ARE YOU PASSIONATE ABOUT?

....

WHAT ARE YOU GOOD AT?

....

WHAT ARE YOU NOT GOOD AT?

....

WHO POSITIVELY INFLUENCES YOU?

....

WHO INSPIRES YOU?

....

....

....

WHAT DO YOU HOPE TO GET OUT OF READING THIS BOOK?

....

....

WHAT SHOULD YOU START?

....

....

WHAT SHOULD YOU CONTINUE?

....

....

WHAT DO YOU WANT TO CHANGE?

....

....

....

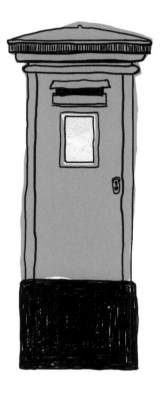

Write a postcard to yourself in the space provided. You can refer back to it at a later date. Write down how you are feeling as you are reading this book. What would you like to get out of it? What help do you need to get? How do you want to help yourself? Don't be scared to write down your hopes and fears.

DEAR ...

USE THESE PAGES AS A SPACE TO JOT DOWN
YOUR THOUGHTS AS YOU READ THIS BOOK

GETTING MENTALLY PREPARED

GETTING MENTALLY PREPARED AND FINDING THE SUPPORT YOU NEED

Being prepared to work doesn't just mean being tooled up with a uniform and equipment to fulfil the role. You also need to adjust your mindset and mentally prepare for this new challenge.

By this I mean you need to have your internal light switched on. Have you consciously decided that you want to change your life by entering employment and contributing to society? Do you want to do something more meaningful with your life? I'm not saying your life at the moment is worthless, but I know that unless you have turned on the switch, you could fail.

In order to switch the light on, you must sort a few things out first. There are many issues that could stand in the way of progress. These might be related to confidence or self-esteem, which you could work through with the appropriate support, or there might be issues of addiction that require professional help. Whatever it is that is standing in your way, I urge you to acknowledge it and do something about it before making a half-hearted commitment that might ultimately be doomed to failure.

To be frank, of course you can start work without switching the light on and dealing with this stuff, but you could be setting yourself up for a fall that would be difficult to recover from.

If you can deal with your issues, you will be ready to flick that switch. This doesn't mean there won't be any challenges ahead of you, but recognition of your own personal problems and the formation of a positive plan will help give you the space to focus your attention on your new opportunity. Your mind will start to fill with a mixture of important new issues as you decide which steps to take next to find a job.

It's important to note that previous issues won't disappear, but they can be compartmentalised in your mind in order for you to move forward. What's vital to understand and accept is that they still need to be dealt with as well as all the new challenges you'll face. If you recognise this, you've got every chance of moving forward with great success.

SUPPORT
(THE KEY TO SUCCESS)

The key or missing link to success is support. The organisation that I founded, Brigade, works with people who have been at risk of or have experienced homelessness to motivate and inspire them to gain meaningful employment. We believe that when someone is committed to changing their life, it is amazing what can be achieved if they are given the right level of support, no matter how vulnerable they might be.

If you were to talk to someone who is deemed a success, such as the managing director of a company, a professional footballer or a chart-topping musician, you would hear that they have people who support them. This allows them to do their work, apply themselves and concentrate.

I believe that if people are to be elevated from a vulnerable position and get back to work, standing on their own two feet, they will need additional support.

RECOGNISING YOU NEED SUPPORT

First of all, recognising that you need support is fundamental. If you don't feel you need it, you won't accept help or ask for it. I believe this is probably the biggest cause of failure.

When I became a single parent, I struggled to balance my career with looking after my son. Getting him to and from school was very difficult, as I needed to be available to go to work. Plenty of people offered to help, but initially I was too proud to recognise that I couldn't do it all. Eventually I did turn to a friend, who now helps me out most days. As a result, I'm happier and so is my son.

BEING OPEN TO SUPPORT

Recognising you need support is one thing, being open to receiving it is quite another. Don't let your pride hold you back. Everybody needs a little help sometimes. Be positive about support – it's a means to an end. It will help you get to where you want to go.

ME

HOME	BENEFITS	SELF-ESTEEM
DEBT	BANKS	HEALTH
WORK	LEGAL	CAREER
FAMILY	EDUCATION	BILLS
FRIENDS	CONFIDENCE	WELL-BEING
	MOTIVATION	COMMUNICATION

Knowing who to go to

In the table above, I've given you a list of some of the areas in life where people often need support. Overleaf you will see a large box with the word 'Me' in it. Using my example, write your own list of all the different areas of life that present you with challenges.

Once you have identified where you need support, then think about who would be best placed to help you. Draw a circle around the areas in which you are already receiving assistance and a box around the ones you where you are not. It's worth putting a lot of effort into this exercise as you'll be surprised by what might come up.

Thinking it through in this way lets you clarify what help you need and in which areas you need it. If you can form a clear picture about the problems you're facing, you will then be able to focus on what to do about them.

ME

HOW TO GET SUPPORT THAT WORKS FOR YOU

When contacting potential sources of support, it's important to understand that they are not just sitting waiting for your call or visit.

- Make an appointment or leave a message with your enquiry or request.

- Don't just assume that someone can help. Be prepared to have to return or to be passed on to another contact.

- Be polite.

- Have a list of questions prepared to avoid forgetting your main points, being nervous or feeling rushed. You can use the note page overleaf to do this.

- Don't be frustrated if the response is not what you need or expect. There might be another solution.

- Before leaving, gather as much information as possible and write it down so that you can refer back to your notes at a later date – in particular, note down contact details for further sources of support.

- Once you are gathering the help and support that you find, map it out against your list and hopefully you can see the improvements you're making as you go along.

USE THIS NOTE PAGE TO JOT DOWN NOTES, QUESTIONS AND
CONTACT DETAILS AS YOU BEGIN FINDING SOURCES OF SUPPORT

_____

_____

_____

_____

_____

_____

_____

_____

_____

_____

_____

_____

DECIDE YOU WANT A

JOB

FINDING A JOB

DECIDE YOU WANT A JOB

I know it's obvious, but it's a big decision. You won't find a new opportunity if you're half-hearted about it, and you're particularly unlikely to find one that you really want and will enjoy.

I am assuming that if you're reading this, you have decided you want to find employment, and meaningful employment at that.

There are lots of online guides to getting a job, all with relevant information. This is my own no-nonsense version. It comes from my own experience over the last 25 years, and I have been lucky enough to always get the job I wanted.

THE BEST WAY TO GET A JOB

There is no doubt that the best way to get a job is to use your own network of contacts. That might include your past or present employer, colleagues, acquaintances, friends and family. If they know you, like you and appreciate your skills, you're halfway there. When looking for a new job, a great reference can make all the difference.

If you don't know of anyone who can help you make connections, there are some simple things you can do to put yourself out there. These require a positive commitment. That means you've got to give 100 per cent to each step in order to succeed. And by success I mean finding full-time employment, earning a living, paying your way in society and striving for a better, happier life. So it's all worth it.

THE SIMPLE THINGS

• Develop your story

• Write a winning CV

• Write a compelling covering letter

• Specific job searching

• Prepare for interviews

To do these things successfully, you need to devote enough time to them and apply yourself with energy, creativity and a positive attitude. You also have to be prepared to make compromises from time to time.

The following points outline what you need to do. I am keeping it brief but, believe me, each of them is vitally important.

DEVELOP YOUR STORY

Everyone has a story. No matter what yours is, I am sure it's compelling and you should tell it. It can include the things you have learned and achieved, the things that didn't go so well, your dreams and what you imagine yourself doing in the future. This is what employers want to hear, and sharing this information about yourself can be a powerful technique when trying to find a new job.

Stories can help you build confidence, increase your own self-awareness and make you memorable to others, so developing and learning how to tell your story are vitally important. It takes work and practice, but making a commitment to improving the quality of your storytelling can help you make a great impact. The story should be part of your CV, feature in your covering letter and should come out during an interview.

HOW TO DEVELOP YOUR STORY

Plot out the chapters in your life – the big things that have helped shape you, the good and the bad. What have you experienced and what have you accomplished? What forced you to learn the hard way?

IDENTIFY THE AHA MOMENTS

The aha moments do not need to be life changing or dramatic, but they must be significant events that helped shape you. One of mine is Sunday lunches around the table with my parents and my brothers. This meant togetherness, and it's really important to me now as an adult and in my career.

REFLECT ON YOUR LIFE EXPERIENCES UP TO NOW

Don't forget to dig deep and think outside the box. Experience isn't restricted to what you studied at school of college, or the jobs you've done so far, though these are also important. There are lots of different life events that can also provide experience that will be valuable in your career.

WHAT ARE THE MOST IMPORTANT EXPERIENCES IN YOUR LIFE TO DATE?

WHAT ARE YOUR GOALS IN LIFE?

REFLECT ON YOUR CAREER SO FAR

WHAT WERE THE MOST POSITIVE THINGS ABOUT THESE EXPERIENCES?

EDUCATION & TRAINING

OUTLINE THE EDUCATION AND TRAINING YOU'VE RECEIVED

HOW DID YOU GET TO WHERE YOU ARE NOW?

WHO HELPED YOU?

WHAT KINDS OF LEARNING/RESPONSIBILITIES HELPED YOU TO IMPROVE YOUR SKILLS?

HAS YOUR MOTIVATION CHANGED?

WHAT WERE THE MOST POSITIVE THINGS ABOUT YOUR EXPERIENCES – FOR EXAMPLE, DID YOU GAIN QUALIFICATIONS OR OTHER REWARDS?

WHAT KINDS OF RESPONSIBILITIES HELPED YOU TO LEARN AND IMPROVE YOUR SKILLS?

HOBBIES & INTERESTS

WORK EXPERIENCE

WHAT HOBBIES OR INTERESTS DO YOU HAVE?

WHAT WORK EXPERIENCE HAVE YOU UNDERTAKEN?

WHY MIGHT AN EMPLOYER BE INTERESTED IN YOUR HOBBIES?

WHAT WERE THE MOST POSITIVE THINGS ABOUT YOUR EXPERIENCE?

WHAT SKILLS HAVE YOU LEARNED?

WHAT ARE THE THEMES IN YOUR STORY?

First identify the key building blocks for your story. This will help you to start writing it.

- In which parts of the story do your
  values and beliefs become apparent?

. . . .
. . . .
. . . .
. . . .
. . . .

- What emerges as hugely important to you?

. . . .
. . . .
. . . .
. . . .
. . . .

- Are you a leader, a motivator, an
  entrepreneur or a fantastic team player?

. . . .
. . . .
. . . .
. . . .
. . . .

QUESTIONS TO ASK YOURSELF:

• How did you get to where you are now?

....

....

....

....

....

....

....

....

....

....

• Who helped you?

....

....

....

....

....

....

....

....

....

....

• Has your motivation changed?

....

....

....

....

....

....

....

....

....

....

....

• What are your goals in life?

....

....

....

....

....

....

....

....

....

....

....

....

....

WRITE YOUR STORY

Come up with a short, sharp version of your story. Read it aloud to yourself and to others. Let them give you feedback. It's a brave thing to do, but you will benefit from it. Try to cut it down to a concise account that is rich in detail but not too wordy.

Once you start telling your story, you will start to understand it more – what the chain of events is and how it's all linked up. You will start to understand the power of your story.

WRITE A 20-SECOND SELF-PITCH

A self-pitch is your 20-second response to the standard interview invitation to 'Tell me about yourself.' If you are not sure what to say, you have not practised enough. You will come across as unconfident and you will feel vulnerable.

Write your self-pitch and practise it with a friend or family member. Mine is on the page opposite.

You can use this self-pitch in whatever way you like. Selling yourself professionally and personally is the best way to impress your potential new employers.

WRITE A WINNING CV

Your CV is hugely important. It is your chance to introduce yourself. Many employers will receive hundreds of CVs for each job advertised. They need to sift through them quickly. In some organisations a computer does the first sift, looking out for key words and mistakes. So coming up with a winning CV is definitely worth taking time over.

Turn to page 44 for my list of the top ten things you need to do to have a winning CV.

I am Simon Boyle and I am a chef.

I have been very lucky in life, as I have been passionate about cooking since I was ten years old. I have been able to apply myself to many different styles of my craft. I love the hospitality industry in general, as it has given me lots of fantastic opportunities.

I am interested in working for your company, as its reputation is impressive.

I think I have some great skills, which I have gained through working for fantastic employers such as The Savoy, Mosimann's, P&O Cruises and Unilever. I could be a great addition to your team and be part of its ongoing development.

I am looking to progress my career and this opportunity seems to fit perfectly.

I live in Surrey with my two children. I love travelling and discovering new things. I have an unhealthy love of 1980s' rock band Marillion. I recently ran the London Marathon and enjoy food photography.

HOW TO BUILD YOUR CV

KEY CHARACTERISTICS

SUMMARY OF EXPERIENCES

SUMMARY OF SKILLS

WHAT MAKES YOU SPECIAL

QUALIFICATIONS

ACHIEVEMENTS

RESPONSIBILITIES

POSITIVE OUTCOMES OF
LEARNING EXPERIENCES

VERY SHORT AND
CAREFULLY
WRITTEN

PERSONAL
PROFILE
1

EDUCATION &
TRAINING
2

3
HOBBIES &
INTERESTS

4
WORK
EXPERIENCE

DUTIES

RESPONSIBILITIES

ACHIEVEMENTS

POSITIVE OUTCOMES OF
WORKING EXPERIENCES

ENTHUSIASMS,
PASSIONS & INTERESTS

SKILLS

ANYTHING ELSE

Your CV will be broken down into the following sections:

YOUR CONTACT DETAILS

Your name, address and a telephone number that an employer can use to get in touch.

YOUR PERSONAL PROFILE

This paragraph gives a description of what you are like and the skills and strengths that you can bring to a position. What do you do best? What are you good at? This can be taken from your personal story.

YOUR EDUCATION AND TRAINING

Briefly outline the time you have spent in school and at college, including the main subjects you covered and listing any qualifications you earned.

YOUR WORK EXPERIENCE

This might include work experience gained while at school and any other part-time or full-time work you have had.

YOUR HOBBIES AND INTERESTS

Employers like to know what you enjoy doing in your spare time, especially any sporting or creative activities. For example: 'I play football for a local team,' or 'I like design and make costumes for plays in my spare time.' This is a chance to show an employer your hidden talents.

REFEREES

Employers will need to know who to contact for a reference.

Over the page is a CV template and an example for you to use. It would be worth your while to type them both up and then save them on a memory stick, so that you can use them as a guide when needed. Fill the template in with the information that you've now gathered. Be creative, think of the skills and experience that make you more employable.

CURRICULUM VITAE

Name:
....
Address:
....
Phone number:
....

Personal profile:
....
....
....
....

Education and training:

Dates: School/College:
....
Qualifications/Achievements:
....
....
....
....

Work experience:

Dates: Company:
....
Your duties/Responsibilities:
....
....
....

Dates: Company:
....
Your duties/Responsibilities:
....
....
....

Hobbies and interests:
....
....
....

References available on request

ANNE JONES

1/23 Rocky Road
London EB 1OZ
Tel: 07XXX XXX XXX

Personal profile

I am a trustworthy and hardworking individual who enjoys a challenge. My current ambition is to become a Motor Vehicle Engineer Apprentice. Practical work on cars is something that has always interested me and led me to gain important experience in my own time. I am motivated, enthusiastic and good at problem solving. I enjoy working as part of a team but can also use my initiative when working alone.

Education and training

2012–present   Rocket Training Ltd   New Ferry

While at Rocket Training, I have been working on my basic skills in literacy, numeracy and ICT. I have also been learning employability skills in order to find a placement to begin my Motor Vehicle Engineering training.

2010–2012   Wallasey High School   Wallasey

I attended Wallasey High School until this year where I learned to work well as a team member through participation in school sports teams. I gained the following GCSEs:

Maths

English

Science

Textiles

Religious Education

Work experience

2014   National Tyres   Liscard

I worked at National Tyres for three weeks to gain important experience in mechanics. While there, I learned a number of new skills such as changing tyres and oil filters. Working as a team, we were able to change and fit exhausts more efficiently.

2013   The Flying Dutchman   New Brighton

During my two weeks work experience, my main duties were to move large amounts of coal, stock shelves and assist customers when needed. My feedback was very positive.

Hobbies and interests

In my spare time I enjoy playing sports, including football and tennis. I played under-16s tennis for my local club, where I won player of the year in 2012. I also enjoy fixing and riding my motorbike and socialising with friends.

Reference available upon request.

KEEP IT REAL

- It must be no more than two pages long.
- Make it punchy and to the point. Try using some of the words on page 46 to describe the work you've done and the experiences you've had. This should help you to avoid falling into the trap of using certain words and phrases repeatedly.

TAILOR IT

- Don't send exactly the same CV to everyone.
- Take the time to change it for each job you apply for.
- Make it relevant to the role advertised; do your research.

INCLUDE A LIFE STATEMENT (SELF-PITCH)

- It's not just about your professional skills; you need to get across a sense of who you are, the person behind the skills. In your life statement, you must get your personality over.
- You can add a bit of interesting personal information, but keep it professional.

DON'T LEAVE GAPS

- If you have periods of time when you were not in employment, be creative when describing them.
- What did you learn during that time? Did you do a course, travel, spend time with family or volunteer?
- Did you learn something about teamwork or communication? Turn it into a positive period of your life.

KEEP IT UP TO DATE

- It's far easier to keep your CV up to date rather than having to start afresh every time you want to look for a new job.
- Add notes to it when relevant; that way you won't forget the important events and achievements.

CHECK FOR MISTAKES

- Don't rush the writing of your CV, especially when you need to get it ready quickly.
- Get someone to check it for you.
- Mistakes will mean it is just thrown out straight away.
- Always use spellcheck.

DON'T LIE

- People think that everyone lies on their CV. It's not true!
- Honesty is vital, and you can always be caught out.
- Employers will check on your background and ask you questions.

BACK UP WHAT YOU SAY

- Make sure you back up what you're saying with some good facts.
- If you say you can do something, then give examples to prove it.

MAKE IT LOOK GOOD

- A CV is an extension of yourself, and image is important in today's world.
- Pretty it up.
- Use bullet points and keep sentences short, snappy and to the point.
- Use a graphics program to allow plenty of space around the paragraphs you write; it will be easier on the eye.

MAKE IT KEYWORD FRIENDLY

- Online search sites and recruiters use keyword recognition. This means they look for certain words relevant to the role they are recruiting for. Make a list of words relevant to the role and use them in your CV. Examples would include: team player, good communicator, challenger, sales performance, KPIs (key performance indicators).

- Have a look at the action verbs over the page to help identify some of these.

**Management/
Leadership Skills**
administered
analysed
appointed
approved
assigned
attained
authorised
chaired
considered
consolidated
contracted
controlled
converted
coordinated
decided
delegated
developed
directed
eliminated
emphasised
enforced
enhanced
established
executed
generated
handled
headed
hired
hosted
improved
incorporated
increased
initiated
inspected
instituted
led
managed
merged
motivated
organised
originated
overhauled
oversaw
planned
presided
prioritised

produced
recommended
reorganised
replaced
restored
reviewed
scheduled
streamlined
strengthened
supervised
terminated

**Communication/
People Skills**
addressed
advertised
arbitrated
arranged
articulated
authored
clarified
collaborated
communicated
composed
condensed
conferred
consulted
contacted
conveyed
convinced
corresponded
debated
defined
described
developed
directed
discussed
drafted
edited
elicited
enlisted
explained
expressed
formulated
furnished
incorporated
influenced
interacted

interpreted
interviewed
involved
joined
judged
lectured
listened
marketed
mediated
moderated
negotiated
observed
outlined
participated
persuaded
presented
promoted
proposed
publicised
reconciled
recruited
referred
reinforced
reported
resolved
responded
solicited
specified
spoke
suggested
summarised
synthesised
translated
wrote

Research Skills
analysed
clarified
collected
compared
conducted
critiqued
detected
determined
diagnosed
evaluated
examined
experimented

explored
extracted
formulated
gathered
identified
inspected
interpreted
interviewed
invented
investigated
located
measured
organised
researched
searched
solved
summarised
surveyed
systematised
tested

Technical Skills
adapted
assembled
built
calculated
computed
conserved
constructed
converted
debugged
designed
determined
developed
engineered
fabricated
fortified
installed
maintained
operated
overhauled
printed
programmed
rectified
regulated
remodelled
repaired
replaced

restored
solved
specialised
standardised
studied
upgraded
utilised

Teaching Skills
adapted
advised
clarified
coached
communicated
conducted
coordinated
critiqued
developed
enabled
encouraged
evaluated
explained
facilitated
focused
guided
individualised
informed
instilled
instructed
motivated
persuaded
set goals
simulated
stimulated
taught
tested
trained
transmitted
tutored

**Financial/
Data Skills**
administered
adjusted
allocated
analysed
appraised
assessed
audited
balanced
calculated
computed

conserved
corrected
determined
developed
estimated
forecasted
managed
marketed
measured
planned
programmed
projected
reconciled
reduced
researched
retrieved

Creative Skills
acted
adapted
began
combined
conceptualised
condensed
created
customised
designed
developed
directed
displayed
drew
entertained
established
fashioned
formulated
founded
illustrated
initiated
instituted
integrated
introduced
invented
modelled
modified
originated
performed
photographed
planned
revised
revitalised
shaped
solved

Helping Skills
adapted
advocated
aided
answered
arranged
assessed
assisted
cared for
clarified
coached
collaborated
contributed
cooperated
counselled
demonstrated
diagnosed
educated
encouraged
ensured
expedited
facilitated
familiarised
furthered
guided
helped
insured
intervened
motivated
provided
referred
rehabilitated
presented
resolved
simplified
supplied
supported
volunteered

**Organisation/
Detail Skills**
approved
arranged
catalogued
categorised
charted
classified
coded
collected
compiled
corresponded
distributed

executed
filed
generated
implemented
incorporated
inspected
logged
maintained
monitored
obtained
operated
ordered
organised
prepared
processed
provided
purchased
recorded
registered
reserved
responded
reviewed
routed
scheduled
screened
set up
submitted
supplied
standardised
systematised
updated
validated
verified

**More Verbs for
Accomplishments**
achieved
completed
expanded
exceeded
improved
pioneered
reduced (losses)
resolved (issues)
restored
spearheaded
succeeded
surpassed
transformed
won

WRITE A COMPELLING COVERING LETTER

A covering letter is often the first thing a potential employer looks at on receiving your application. It's not an alternative but an essential accompaniment to a CV, and gives you the chance to highlight specific points about yourself that are relevant to the position.

Once again, here are my ten top tips for getting it right:

1 ALWAYS SEND ONE

A covering letter is a way to introduce yourself in a way that will get you noticed.

It explains who you are and why you are getting in contact with the employer.

It lists your contact details so it's easy for a prospective employer to get in touch.

2 DON'T REWRITE YOUR CV

The letter should include only the most relevant and interesting highlights from your CV.

3 SELL THE SIZZLE

Don't waffle on. Keep it concise and relevant to this particular job.

Make the reason for your application clear.

Sell yourself.

Finish with a call to action – ask that they invite you for a meeting or interview, or let them know you will be back in touch to discuss your application.

4 TALK ABOUT THEIR COMPANY

Research the company or organisation.

Include some information that displays your knowledge.

Tell them why you are attracted to them specifically.

5   TELL THEM WHY YOU ARE RIGHT FOR THEM

Pick out the qualities they mention in their advert and make sure you have mentioned them in your CV.

Briefly mention them again in your covering letter, providing good examples that can be backed up. For example: 'You will see I match your requirements perfectly. I have worked in the hospitality industry for five years and have developed my key customer-service skills. I am friendly, approachable and professional.'

The covering letter is then an additional selling tool, reinforcing why you are right for the job.

6   REFLECT YOUR PERSONALITY

Make sure your letter shows you are motivated and enthusiastic.

Don't include any negative comments and try to give an impression of who you really are. Maybe you're a people person, have a strong sense of humour, or you're a great teacher, supporter, analyser or collaborator. Identify your relevant strengths and be sure to include them.

7   KEEP IT BRIEF

A good letter should draw the recruiter's eye to the relevant experience on your CV.

It should take up no more than one page and it should provide a sense of who you are.

If it's too long and doesn't get to the point, you will not gain a second glance.

8   SEND IT TO THE RIGHT PERSON

Whenever possible, send your letter to a named and titled person. This is particularly important if you are making a general enquiry regarding whether any positions might be available rather than replying to a specific job advertisement.

If this is the case, do your research to find the relevant person to write to – look online, call reception desks, ask around. Getting the most appropriate person to open your letter can help dramatically. This is much better than making a 'Dear Sir or Madam' approach.

9 MAKE SURE YOU SIGN THE LETTER

Unless you need to sign an accompanying form, your covering letter is the only place you must provide a signature. Take a moment before you do. It's still a very strong signal of your authenticity.

10 MAKE IT PRESENTABLE

Make sure your covering letter looks good, is clearly laid out with no typos and is easy to read. You should check for spelling errors two or three times.

Get someone to check it for you.

Here's an example of a template and a covering letter that you can use. Practice makes perfect, so find a job advert in the field you'd like to work in and try writing a covering letter. Use the checklist to remind you of all the main points you should cover. Set it aside for a while and then review your letter. Spend a few minutes reading through what you have written and see if you can make any improvements. If possible, ask someone else to read the letter and give you some feedback. When you read their comments, think about what you have done well and what you could improve on.

YOUR NAME ....
YOUR ADDRESS ....
TEL .... EMAIL ....

....
.... NAME AND ADDRESS OF
.... WHO YOU ARE WRITING TO
....
....
....

DATE ....

Dear ....

I am writing ....

....

....

....

....

....

....

Yours sincerely ....

SIGNATURE ....

NAME ....

....

John Doe

Address: 24 Mansfield Drive, Cheadle, Manchester, M23 4DJ
Email: johndoe@example.com   tel: (01XXX) XXX XXXX

Catherine Jenkins
Personnel Manager
Oldbury Foods Limited
Manchester
M2 3LL

THE NAME AND
ADDRESS OF WHO
YOUR LETTER
IS TO

YOUR CONTACT
DETAILS AT THE
'HEAD' OF THE
LETTER

3 July 2014   DATE BELOW ADDRESS

Dear Ms Jenkins

INCLUDE THE SURNAME OF THE PERSON YOU
ARE WRITING TO, WITH 'MR' OR 'MS'. IF
UNSURE WHO, USE 'DEAR SIR OR MADAM'

I am writing in response to the advert in your restaurant window, looking for a catering assistant in your restaurant who can start in September 2016.

In your job description, you said you are looking for someone with an interest in food who can deal with a fast-paced environment. As a hard-working individual who is enthuasistic about the catering industry, I believe I have the personality and skills that you are looking for. I have previously worked in a busy restaurant, helping to prepare and serve food, clear tables and make sure that the customers have a positive experience.

Please find enclosed my CV, with some more details of my background. Feel free to get in touch if you would like any further information, or to arrange a meeting. I look forward to hearing from you.

Yours sincerely

SIGN YOUR LETTER BY HAND.
INCLUDE YOUR NAME TYPED
BELOW YOUR SIGNATURE

John Doe

SEARCHING FOR A JOB

ONLINE

There are four main ways to search for a job:

There are many sites you can use to search for jobs. One example is www.gov.uk/jobsearch. You can search here for jobs according to type and location.

ADVERTS IN NEWSPAPERS, JOB CENTRES AND TRADE MAGAZINES

Jobs are advertised in a wide range of publications from national dailies like *The Times* and the *Mirror*, to local and regional daily or weekly newspapers. You can search for regional newspapers on the News Media Association's website www.newsmediauk.org

Job centres offer the following facilities:

- Job boards with vacancy details on them
- Jobpoints – touch-screen computer terminals you can use to search for vacancies
- Jobseeker Direct (tel: 0845 60 60 234) – a phone service you can call to get details of job vacancies, request application forms and arrange interviews
- Jobcentre Plus website

RECRUITMENT AGENCY

Recruitment agencies can take the hard work out of job searching, especially if you have skills that local employers are looking for. Agencies can specialise in temporary work, permanent work or specific sectors. Look at https://nationalcareerservice.direct.gov.uk/advice/getajob/howtofindajob/Pages/recruitmentagencies.aspx to find out how to deal with recruitment agencies to make sure they're working hard to find you the right job.

NETWORKING

Networking is using word of mouth to find out about the many jobs that aren't advertised by recriutment agencies. If you learn to network well, you'll get inside information on jobs and careers, and build a list of contacts who can help you find work. Applying for jobs that aren't advertised also cuts down on the amount of competition for each vacancy.

THINK ABOUT EACH OF THE CATEGORIES
AND TRY TO ANSWER THE FOLLOWING QUESTIONS:

WHAT IS GOOD/EASY ABOUT APPLYING FOR A JOB IN THIS WAY?	WHAT CAN BE DIFFICULT/CHALLENGING ABOUT APPLYING FOR A JOB IN THIS WAY?

ONLINE JOB HUNTING

ADVERTS IN NEWSPAPERS/ JOB CENTRES/TRADE MAGAZINES

RECRUITMENT AGENCIES

NETWORKING

THE MUST-DOS WHEN JOB HUNTING

BE PREPARED

Have a voicemail system in place.

Have your phone topped up so you can receive messages.

Respond quickly to any calls or messages.

Have your up-to-date CV ready – even if you're not looking for a job, you never know when an opportunity may come up.

You may wish to consider a LinkedIn profile and other social media sites.

LinkedIn is a business-oriented social media site and is used to help people network to find job opportunities. See page 56 for social media advice.

GET HELP

Use any free career-advice service open to you.

Colleges, job centres and public libraries often hold workshops or classes and have computers and printers plus other resources available, and these are often free or inexpensive to use.

CV AND COVERING LETTER TEMPLATES

Have your CV and covering letter templates ready and then change them for each job application to fit the company or job description.

Show explicitly how you can match any of the requirements.

The contact information and opening and closing paragraphs can often stay the same.

Don't forget to use spellcheck and ask someone to review your CV and letter for you. Double check them again yourself before you send them.

USE THE INTERNET AND SEARCH ENGINES

The Internet is the fastest way to look for a job.

Job search engines are like online noticeboards. They change frequently. Use the various job search engines, company sites, associations and other sites with job postings to find any positions that might suit you.

Sign up for job alerts and you will receive job listings by email. All major job search engines and websites have this facility.

Don't limit yourself to the main job-searching websites: Also go to smaller, more niche sites that focus on a particular industry, location or career field.

REFERENCES

It is important to have your references ready.

Ask your referees in advance and make sure you have their details correct on your CV. You will need their name, job title, company, address, phone number and email address.

USE YOUR OWN NETWORK

Remember it is always easiest to get a job through personal recommendation.

Tell everyone you know that you are looking for a job and ask if they can help.

USE SOCIAL NETWORKING

Social networking is fast becoming the most important way to communicate and promote yourself in the employment market. But be very careful. While you can use these sites to gather information about individuals and companies you'd like to work for, they can do the same with regard to potential employees. Once something is posted on the Internet, it's very hard to get rid of it. So think twice about posting pictures or news that you wouldn't want an employer to see. It really can affect your chances.

These are some of the biggest sites you can join:

TWITTER

Follow prospective employers or companies that you'd like to work for. This is often the quickest way to keep up to date with their news, and job vacancies are frequently mentioned.

FACEBOOK

Unlike Twitter, you can view whole pages of information about companies you are interested in as well as keeping up with their latest news.

LINKEDIN

This is like Facebook for professionals. People use it for connecting and building a network of business contacts.

BE OLD SCHOOL

Walk the street and drop your CV in to places where you'd like to work.

Make an appointment with the HR manager or the person you think will be interested to meet you.

You'll be surprised by the reception you might get. Don't forget, they can save money by hiring you directly.

USE HIGH-STREET JOB AGENCIES

Using high-street agencies provides you with a friendly face to interact with.

Temp agencies can be particularly helpful. Sometimes temporary jobs lead to full-time work.

PREPARING FOR INTERVIEWS

What are the things you need to think about when preparing for an interview?

Look at each of the questions in the boxes on the next page and note down some answers. To help you with the final box 'What else should you think about?', answers here would include: 'Prepare to be asked about the experiences you've included on your CV,' or 'Bring a copy of your CV to the interview.'

When you've written something in each box, ask yourself which area you feel most nervous about, then think of ways you can prepare to make this category seem less daunting. Ask others what they would do to prepare.

PLAN AS FAR IN ADVANCE AS YOU CAN

Work on the common questions you think the interviewer will ask. For example, 'Tell me about yourself'; 'Can you take me through your CV?'

Work on two-minute responses that you feel comfortable to give as replies.

Think about any challenges they could throw at you, such as asking about gaps in your CV. Think of answers that will let you present yourself in a positive light and help them to understand why there is a gap, or tell them about what you learned during that period of unemployment. Don't panic – plenty of people have gaps and no one's career history is perfect. Having an answer prepared and practised will stop you coming across as defensive.

WHAT QUESTIONS MIGHT YOU BE ASKED?

WHAT SHOULD YOUR BODY LANGUAGE BE LIKE?

WHAT SHOULD YOU DO THE NIGHT BEFORE/MORNING BEFORE SETTING OFF?

WHAT ELSE SHOULD YOU THINK ABOUT?

JOB DESCRIPTION AND PERSON SPECIFICATION

When you applied for the job, there might well have been a job description and person specification that formed part of the longer job advertisement. You will have had to take notice of this to tailor your CV and covering letter appropriately. Now that you have an interview, it's essential to go through them again and make sure you can show how you match up.

A job description is a list of the tasks, functions and responsibilities of a position. It might give you information such as who you would report to, what qualifications and skills are needed, and a salary range. It might include a list of competencies such as team requirements, planning and organisation. A job description is a useful tool and should be read and understood fully. Make sure you can give examples from your past experience that show how you will be able to fulfil the role.

The person specification is a profile of the candidate required for the work. It might contain educational requirements, previous experience, specialised skills, interests, personality and physical requirements. Make sure you can show you have the skills, knowledge, experience and right attitude for the job.

When preparing for the interview, make a list of the ways in which you match up to the job description and person specification, and spend time practising articulating each one in an interview. Record yourself and listen back to make sure you're coming across as you would wish.

TRICKY QUESTIONS

So many people stumble into interviews and are thrown by the questions that they are asked. But most interview questions should be expected and planned for, and the answers rehearsed. Take the time to think about potential challenges that could come up in an interview. Look at the list below, study it and learn to deliver a quality answer with confidence:

1. WHAT ARE YOUR WEAKNESSES?

The best way to answer this dreaded question is by mentioning one or two minor faults and then talking about your strengths. Stay away from personal flaws and talk about professional traits: 'I am always looking to improve my communication skills and recently joined a writing course, which I am finding very useful.'

2. WHY SHOULD WE HIRE YOU?

Summarise your experiences into one main point: 'With XX years' experience working in the hospitality industry, I could be a great asset to your business. I am confident that I would be a great addition to your team.'

3. WHY DO YOU WANT TO WORK HERE?

Here, the interviewer wants to hear that you have given this some thought and that you have not just sent your CV to lots of different employers, hoping that something would turn up. Be specific: 'I selected key companies whose mission statements are in line with my values and beliefs. I get excited when I see what your company has achieved and I think I could be very happy working for you.'

4. WHAT ARE YOUR GOALS?

It's best to talk about short- and medium-term goals. Don't talk yourself out of an opportunity now because of a pipe dream that might never happen. 'My immediate goal is to find a job with a growth-oriented company. My medium-term goal is to work hard and grow into a position of responsibility.'

5. WHY DID YOU LEAVE/ ARE YOU LEAVING YOUR JOB?

If you are unemployed, try to put a positive spin on the situation: 'I have managed to keep learning despite losing my job because of staff reductions in the workforce.'

6. WHEN WERE YOU MOST SATISFIED IN YOUR JOB?

The interviewer wants to hear what excites you, what motivates you. Give them examples of times when you felt fantastic about your job. Talk about personal achievements at work, things that were important to you and the company.

7. WHAT CAN YOU DO THAT OTHER CANDIDATES CAN'T?

What makes you unique? This can be an assessment of your personal and professional experiences, skills and traits. Be concise. 'I am unique because I have a strong set of skills, I love building relationships with customers and I like to work harder than most to get the job done properly, on time and on budget.'

8. WHAT ARE THREE POSITIVE THINGS YOUR LAST BOSS WOULD SAY ABOUT YOU?

Think about what has been said before. Don't brag but be honest. 'My boss has said that I have a great eye for detail, that he can rely on me to get the job done with real creative flair and that he has enjoyed me being a part of his team.'

9. WHAT SALARY ARE YOU EXPECTING?

It is usually best for them to tell you the range first. But you can prepare yourself by researching the going rate for that position. You can say, 'I am sure that when the time comes we can agree on a reasonable amount. In what range would you typically pay someone with my level of experience?'

10. IF YOU WERE AN ANIMAL, WHAT ANIMAL WOULD YOU BE?

Interviewers want to see how quickly you can answer this question. If you answer 'a bunny', you will be seen as a soft, passive person. If you answer 'a lion', you may be seen as aggressive. What type of personality would it take to get the job done? What impression would you want to make?

FORMAT OF THE INTERVIEW

If it's not obvious, find out the format of the interview, as this will help you to prepare. How many people will be involved? Where will it be held? How formal will it be?

RESEARCH THE COMPANY

Find out about the company and any organisation behind it. Look for news stories you could mention. Review their website and any plans they have on the Internet. Ask your own friends and family if they know anything about the company. You'll be surprised how far a little extra information can take you.

BEFORE THE INTERVIEW

Looking your best will give you confidence and boost your professional image.

- Get your outfit ready.

- Shine your shoes.

- Get a haircut.

Plan where you are going, check travel times and your transport options.

Find a coffee shop close by and give yourself time for a drink and a toilet stop before you go in. This will reduce your anxiety.

Take a copy of the job description, the job specification and your CV in a portfolio case. Read them through prior to entering.

TRY BEFORE YOU BUY

Visit the employer and if possible experience the business as a customer. It will give you an insight into who they are and if you'd like to work with them. It might also give you some good pointers for questions to ask during the interview.

DURING THE INTERVIEW

Remember it's a two-way conversation and ask questions, such as 'How do you find working here?'

Don't be afraid to take a breath or pause before answering a question. Acknowledge that you need a moment for consideration before answering a tricky question. Slow down the conversation if you are answering too quickly. This happens when you're nervous. It's your interview as well, so you can also control the flow.

BE CLEAR OF ANY NEXT STEPS

Don't be afraid to say: 'So will I hear from you by Monday?'; 'If I don't hear by Wednesday is it OK to drop you a line?'

AFTER THE INTERVIEW

Find a quiet place and write down as many of the questions that you were asked as you can. Rank your answers 1–10 in terms of how well you think you did. This will help you for your next interview should you need one.

The next day, drop your interviewer a line thanking them for the opportunity and asking them any outstanding questions.

Don't stalk them. Sometimes the company has a strict 'Don't contact us' policy. But, if not, follow up if you don't hear from them.

Even if you are not successful in securing the role, remember it has been a good opportunity to engage and grow your own network. And it has also been an opportunity to practise your interview technique.

If you are successful, well done! Now it's time to start thinking about either second-round interviews or starting in the role.

USE THIS PAGE TO PRACTISE ANY TRICKY QUESTIONS YOU ARE PARTICULARLY NERVOUS ABOUT

BE PREPARED: HOW TO TACKLE AND JUGGLE FINANCES

WHERE TO BEGIN

If you've been out of work for a while, you may well be receiving benefits. Understanding whether you're still entitled to these is important. What gaps will you have if you switch from being supported by the state to being employed?

I have found that there can be a funding gap of quite a few weeks between the day your benefits stop and when you receive your first pay packet. This can be stressful and hard to manage, and might leave you short of money to pay rent, debts, bills, buy food, etc.

I am sure if you explain your position to your job-centre advisor, then you could be supported through this period. It's tough, though, and very often puts people off going back to work. My advice is to find out what government help is currently available to see you through this process. It often changes, so it's worth asking what support is on offer at the moment or if there is something that might be available in the near future. Ask about Back to Work grants; these can often help with uniform, travel, etc.

The paperwork involved in making these changes can be challenging – there is a lot of information to be absorbed as well as forms to be filled in and questions to be answered. Get someone to help you, as sharing the load will make it much easier. If you approach the process calmly, you can often negotiate an overlap in your benefits. Housing benefit is one area where more support seems to be available. For example, if you're enrolled on an apprenticeship scheme and receiving an apprenticeship salary, it is often the case that you can keep your housing benefit. It's well worth looking into.

This is hard, but I would also advise that you try to reserve some funds as you approach your employment date. You will incur extra financial costs such as travel to and from your place of work, and there's no excuse for not being on time and ready to work. Travel money can be a huge barrier to employment. Anything you're able to put aside will help bridge the gap. Some hard choices need to be made in order to have some spare cash. I've included a Budget Planner on page 78 that you can use to help you with this. You can type this into an Excel spreadsheet as this will allow you to update and make changes.

ME AND MONEY

I don't know about you, but I am a nightmare with money.

I recently started to get my head around money, in particular my attitude towards it and how I handle it. Through doing this, I started to value it much more wisely.

In this chapter I hope that I can help you start to prioritise money as an important aspect of your life and how to deal with it much better. If you consider and practically use my advice you should feel much better, more controlled and know exactly where you are in terms of your finances.

THINK FOR A SECOND

I know it sounds like a stupid question to you, but is money a priority to you?

Is it a big priority or less of a priority in your life?

. . . .

. . . .

. . . .

Why do you think it is more or less of a priority?

. . . .

. . . .

. . . .

Answer the questions below:

Is borrowing money a good or bad thing?

Is it worth saving if you have nothing you want?

Money helps us to enjoy life. So should you spend it when you have it?

The best things in life are free

If you look after the pennies the pounds will look after themselves

Money doesn't grow on trees

Where there is muck there is brass

Money is the root of all evil

Money makes money

Neither a borrower nor a lender be

Riches have wings

You can't take it with you when you die

Do you know what these mean?

. . . .

. . . .

Have you every been in a situation where one of these has rung true?

. . . .

. . . .

Do you know of any other sayings about money?

. . . .

. . . .

Has your attitude to money changed over the years?

What, how and when did this happen?

. . . .

. . . .

Our attitude towards money can be shaped by many things.

My attitude to money and making sure I earn it honestly and try my best to look after it comes from my parents. They seemed to look after theirs. They weren't wealthy; in fact my dad was hard working but was made redundant several times, having to move frequently and further south each time. Therefore their money needed to be looked after. Even though I understand that they looked after their money, I have to admit that I haven't always done the same.

Many things influence our attitudes to money:

- Our parents
- Our family traditions
- Our religion or culture
- Our personalities
- Our friends
- Our ambitions and aspirations
- How much money we actually have

Depending on how the above issues affect us; it will shape the way we value and look after our money.

Our attitudes do change as we grow older; things become more or less important as life's experiences take their hold.

For example, if you have borrowed money to buy something and then found it really hard to pay it back, this makes you think twice the next time you want something. Having said that, not everyone does, and this is how debts can build up. Putting stuff on credit cards is dangerous and you can get yourself into huge trouble if you are not careful.

Thinking and knowing about what you have to spend and what you don't is a vital part of money management. This needs to be really thought about. This is called budgeting.

BUDGET

Now I can say that I thought I knew what came in and out of my bank each month. But honestly I didn't. I often get worried about money, and rather than work it out and truly know, I would put bills and bank statements in a pile and try my best to forget about them. Do you do that?

This is crazy, and by dealing with them you solve half the problems that ignoring them can bring.

Think of budgeting as organising your socks, putting them into pairs and placing them in a drawer so that you know how many pairs you have and when you might need to wash some or even if you need to pull some out as you are going away for a holiday or a break.

When your socks are not paired and just thrown in the drawer, when you need them you can never find them. When you are desperate, the only ones that match have holes in. It's so frustrating. This would have been easily solvable if only you had given it a little bit more time.

This is really all about placing some importance on the financial implications of not spending the time to sort stuff out. It's a personal choice ultimately and one that should be taken seriously.

You'll feel so much better if you do. If not immediately, you will eventually.

LEARN TO BUDGET

A budget is a document that shows the money coming in and everyday spending. Working this out can really help you understand what money you have and don't have to spend.

Firstly, you need to establish what you are spending your money on.

Use the personal budget planner to help you do this. See page 78.

You can add some extra items of expenditure if there are not enough categories in the table.

PERSONAL BUDGET PLANNER

INCOME

TOTAL INCOME

EXPENDITURE

TOTAL EXPENDITURE

OUTSTANDING DEBT

BALANCE

INCOME – EXPENDITURE

ADVICE

To work out your income, think about any money you receive regularly, such as:

- An allowance
- Jobs that pay full- or part-time
- Benefits from the government

To work out your expenditure, think about any money you spend regularly, such as:

- Rent
- Bills
- Food
- Travel
- Clothes
- Entertainment (music, books, magazines, cinema tickets)
- Gym
- Loans and debts – don't forget them as they need to be paid as a priority

Look at the balance. Have you got a positive or negative number?

If it is negative, what do you think can be done?

. . . .

. . . .

. . . .

Are there areas of spending that can be reduced so that you can make some savings?

. . . .

. . . .

. . . .

Is there anything unexpected that may affect your budget?

. . . .

. . . .

. . . .

Is there something that you need or want that would affect your budget?

. . . .

. . . .

. . . .

Can you seek some advice on concerns, solutions or ideas that you may not be able to see?

. . . .

. . . .

. . . .

Other places to seek money advice can be found at www.moneyadviceservice.org.uk

How does completing a budget help to avoid financial problems and achieve savings goals?

. . . .

. . . .

. . . .

KEEPING TRACK OF YOUR BANK ACCOUNT

There are a couple of really good reasons why you need to keep track of your bank account.

Firstly, you don't want your financial information getting into the wrong hands.

Secondly, you need to understand how to keep an eye on your account by reading your bank statement.

Now I know it's all boring stuff. The last thing you want to do is look at all this bank stuff. But it is much simpler than you think, and, once you start it, it really is easier and not that time consuming.

KEEP YOUR INFORMATION SAFE

Most people are a bit blasé about their PINs (personal identification numbers). They are all the same for each bank account and store card they own. Careless chitter-chatter can lead to your information being leaked.

SPENDING FREELY

No one has the right to tell you what you can and can't do. However, sometimes the people around you do know best. 'One thing is for sure, **NOT** understanding how much money you have to spend is a huge mistake.'

One easy way to fully understand your day-to-day cash flow is to look at your bank statement. I know, yawn yawn!

BANK STATEMENTS

Where do you keep your bank statements?

It is best to open them straight away and file them once you have checked the information. This means they are easily accessible and you can refer to them as often as you need to.

Check your statement regularly.

By checking your bank statement regularly you can see if there have been any errors such as double purchases or over-changing.

Maybe someone has access to your account that shouldn't have?

By checking each statement you can make sure that only you are spending your money and that you are only paying for the things you have bought.

Bank statements are a really useful tool for managing your money. I know they are not the most exciting things to read but they will show you:

- How much money went into your account during a month. These transactions are called credits.

- How much money left your account during the month. These transactions are called debits.

- Where that money went to, i.e. the place/people who took your payment.

- Any interest or bank charges made on your account. This is either an agreed charge or a special charge for spending unauthorised amounts.

I don't know if you are like me, but these days I hardly use cash. Most of what I buy is online. That's why I find using bank statements is necessary. Otherwise we really are unaware of what we are spending. Looking at the statement means we can check the right amount has left or come into the account and gone to the right person.

If you think the wrong payment has left the account or gone to the wrong person, contact your bank immediately so they can investigate and stop any other payments going out while they do so.

Your information on your bank statement is not only useful to you, but also to criminals, so be careful where you keep them so that they don't fall into the wrong hands. Identity theft is very common; a criminal doesn't need to look like you to steal from you.

HOW TO READ A BANK STATEMENT

Do you really know how to read a bank statement? I didn't for ages.

10 High Street Hackney London RB1 5TL

Account Name Mr No One

Statement period: 1 June to 30 June
Account number: 64748693

Your Bank Account Details

DATE	PAYMENT TYPE	DETAILS	PAID OUT
01-Apr		BALANCE BROUGHT FORWARD	
02-Apr	ATM	Cash ATM Tower Bridge Rd	20.00
03-Apr	DR	Supermarket Putney Card Transaction	50.00
05-Apr	DR	Birthday Card Shop	3.79
08-Apr	DD	South Eastern Water	25.00
		Bank Charge	2.50
09-Apr	CR	Salary	
12-Apr	SO	Bank Loan	60.00
15-Apr	CHQ	123	150.00
17-Apr	DR	Long Field Chemist	5.60
17-Apr	CR	Branch Pay In (CHQ)	
20-Apr	ATM	Cash Dispenser Camden	30.00
24-Apr	DD	ZZ Mobile Phones	21.83
28-Apr	DR	Bank Card Charge	1.75
30-Apr	ATM	Cash Dispenser	20.00

ATM hole in the wall machine

DR a payment through a purchase

DD a set up direct debit

SO a standing order that you will have agreed to

CHQ a cheque that you will have written and given to someone

Sort Code: 00-00-00

PAID IN	BALANCE	
		Balance is what money you have at that particular time
	96.00CR	CR means you are in credit
	76.00CR	
	26.00CR	
	22.21CR	
	2.79OD	OD overdrawn
	5.29OD	Bank charges can apply if you have not got prior agreement
450.00	444.71CR	
	384.71CR	
	234.71CR	
	229.11CR	
55.00	284.11CR	
	254.11CR	
	232.28CR	
	230.53CR	
	210.53CR	

NB - some banks do use slightly different abbreviations but there should always be some kind of text reference to help you to identify what the record refers to

HOW TO KEEP SECURE:

- Use a PIN that no one else knows, make it difficult to guess and change it if you think someone might know it.
- Use a different PIN for each card you have.
- Never give your PIN to someone else.
- Your bank will never ask for your PIN, so if they do be very cautious.
- Always shred or destroy documents that contain personal details such as your name and address on financial documents that you don't need to file.
- Consider having electronic statements rather than printed ones, but still check them!

PERSONAL BUDGET PLANNER

	monthly budget	monthly ACTUAL	(difference)
gross income +benefits			
wages paid			
tax			
national insurance			
net income			
benefits			

FIXED EXPENSES	monthly budget	monthly ACTUAL	(difference)
rent / mortgage			
home insurance			
electricity			
gas			
home maintenance			
water			
work travel			
car insurance			
fuel			
maintenance			
other			
SUB-TOTAL			

	monthly budget	monthly ACTUAL	(difference)
VARIABLE EXPENSES			
TV			
home telephone			
mobile			
out-of-work travel			
childcare			
child support			
food shopping			
toiletries			
clothing			
eating out			
gifts			
going out			
SUB-TOTAL			

FINANCE PAYMENTS	monthly budget	monthly ACTUAL	(difference)
personal loan 1			
personal loan 2			
credit card 1			
credit card 2			
savings			
SUB-TOTAL			

TOTAL EXPENSES			
TOTAL INCOME LESS EXPENSES			

TRAVEL

As well as the cost of travel, you also need to work out the best way to get to work. Things to consider include shift patterns and the number of hours you are required to work. When you're applying for a job, remember to consider how easy it will be to get there, and investigate train, Underground and bus services during the week and at the weekend if you might have to work then.

It might not seem important now, but when you're tired and stressed the journey can be an important factor in determining whether you turn up or not. If possible, it's best to work within an hour from where you live. Also look for somewhere that has at least two methods of transport to and from it. This means that if there is a problem with one link, there is another option that can get you to work.

Don't forget that travel delays happen; they are part of daily life. This means that you can't continually use them as an excuse. You simply need to allow enough time to overcome those delays. Make sure you check the travel situation before leaving or leave earlier to ensure you arrive in good time in spite of any delays. Arriving early is good for you anyway, as you will be mentally alert as well as physically. Being late, rushing and starting on the back foot is never an ideal beginning to a good day's work.

UNIFORM/ EQUIPMENT NEEDS

Most organisations should equip you with any necessary equipment – tools and uniform – but don't take it for granted. For instance, chefs usually have their own knives. Make sure you know what's required before you turn up, and check what's included in the job offer.

EATING

Depending on the role, hours worked and the type of employer, you might need to supply your own food at break and mealtimes. Most hospitality businesses supply food, but offices don't tend to. If you rely on convenience foods such as sandwiches from a local shop, this can take a big chunk out of your wages. Use your own cooking skills to produce something cheaper. Homemade usually means tastier, more wholesome and slower burning. In other words, it's far better for you and will keep you from getting hungry too quickly. The food you eat will directly affect your performance. Think of a racing car – it requires the correct high-quality fuel, longer-lasting tyres plus the skill of the driver to ensure a winning performance.

Plan your meals so you're not eating on the hoof. If you don't, you'll end up grabbing rubbish that is unsatisfying and bad for you.

Prioritise eating over other indulgences such as smoking, drinking and eating sweets. You can't perform well at work unless you are well nourished. If you can't afford food, don't go hungry. Find a community food bank – enquire at your job centre or local council offices. You may find that you need to be referred by a caseworker, charity or job centre.

STAYING HEALTHY

Being physically capable of work is obviously highly important. This means getting plenty of sleep, exercise and downtime to relax and think.

Sleep in particular is often undervalued. It is well worth giving yourself plenty of chill-out time prior to going to bed. Eat a light meal at least four hours before bed and get changed into your nightwear at least an hour before. Read a book and think of something other than work. You should sleep at least eight hours when possible. This will have a dramatic effect on your ability to think, act and focus, not to mention how you deal with and manage problems and challenges at work as well as in your personal time.

DO YOUR HOMEWORK

Preparing yourself intellectually doesn't mean that you need to go and read an encyclopaedia. It means you should read up on the company to get an understanding of the type of business it is and how it operates. What products does it deliver and how? Who does it work with and why? If you can get to grips with these fundamental issues, you'll feel well prepared and your colleagues will be impressed. Once you actually start in the role, your employer will often help you to learn more about the business.

IT'S MASSIVELY
IMPORTANT
THAT YOU TAKE
YOUR NEW
RESPONSIBILITIES
SERIOUSLY

TAKING RESPONSIBILITY AND ACTING LIKE A PROFESSIONAL

Now that you've been successful in finding a job, the next step is to take your new role seriously. The organisation is putting its trust in you, so take some time beforehand to think through what will be expected of you and what is at stake. It's somebody's business you're working in, and shareholders, stakeholders and other staff members all have a collective mission to achieve certain results that will secure everyone's position within the company or institution. Anyone who jeopardises this will not fit in and eventually be managed out. It sounds tough, but that's life.

Being responsible has two main definitions:

- Having an obligation to do something, or having control over or caring for someone as part of one's job or role.

- Being the primary cause of something and so able to be blamed or credited for it.

It's massively important that you take your new responsibilities seriously. If you do, you will fulfil your new role with real enthusiasm and do a great job. You will apply yourself well. You will learn what's needed and what's not. Once you take responsibility for yourself and your work, people will think you are capable, reliable and able to take on more, which could lead to promotion.

BEHAVING PROFESSIONALLY

Part of taking responsibility means making sure you've grasped the basic meaning of professionalism. This is something that is very often taken for granted but, when you really think about it, who teaches professionalism? Parents, schools, colleges, job centres?

Well, yes, they should, but, very often I meet people who claim to be professionals and yet they have missed the very basics of what this means. They may not dress professionally or act or speak as a professional.

Personal issues can be a barrier to success. It is so easy to let your personal life and challenges get in the way of your working responsibilities.

You simply need to leave these at the door whenever possible. If this is not possible, let someone know what's going on and then park it until you have finished your duties. This is hard to do, as your mind is full of feelings and emotions. But it is vital to concentrate on what you're being paid to do. If you have a company that is willing to listen, support and even assist you, that's great. Don't abuse people's goodwill, though; they can tire and become frustrated if you're not careful.

Behaving professionally can refer to so many areas, but basic timekeeping, and looking and acting the part, are vital. If you are a team player and perform your duties with care and a professional attitude at all times, your employer will respect your commitment.

TIME MANAGEMENT

Time management is very important and probably the hardest thing to get right. So, what does time management mean? Well, it's about getting to work and appointments, achieving tasks, and objectives and so many others things *on time*.

How do you achieve good time management? By preparing, well in advance, what you want to achieve and by when.

So, for example, if you are to start work at nine o'clock in the morning:

Clear the day of personal appointments (even if that appointment is just an arrangement to phone someone or meet up for a chat).

Make sure your diary is blocked off for work.

The night before, make any preparations that will help you in the morning – get your clothes ready, make an advance purchase of your travel tickets, put petrol in the car, etc.

Check what time you need to leave, remembering to give yourself extra leeway in case there's a delay en route.

Set your alarm well in advance of when you need to leave.

Get a good night's rest.

In the morning, as soon as you can, check for travel delays or cancellations. If there are, then you'll need to get moving!

Shower, dress and have breakfast in a speedy fashion, setting the tone for the day.

Leave and hopefully you'll arrive on time with 15 minutes to prepare before your shift.

BENEFITS OF GOOD TIME MANAGEMENT

People will learn to trust and respect you when you're on time each and every day.

You will start to gain confidence and self-esteem from being on time.

Your productivity will improve.

You will start to set higher goals for yourself.

You will have the ability to take on new challenges.

So, if you get time management right,

YOUR WHOLE LIFE WILL CHANGE.

I have decided to add this section as, without wanting to insult any readers, I feel that there are many people who have not been shown how to perform some basic life-skill tasks that will help them look like a professional person. I know that some people will know this stuff, and you will pass over it. But many people don't, so it is important to set it out here. It will be of benefit to you every day in the most basic way, and I'm glad someone taught me how to do this stuff. A friend once said to me, 'If it looks like a duck and walks like a duck, it must be a duck.' If you look like a professional and act like a professional, people will think that you are a professional. It's that simple.

LOOKING AND ACTING LIKE A PROFESSIONAL

I have also always thought that dressing smartly puts you in a professional frame of mind as you set off for work. I have known people to dress smartly despite the fact they get changed into a uniform after arriving at work. Although this seems a bit drastic, the principle is a good one. I once knew someone who dressed for work in a suit, although his role did not require it of him. His enthusiasm got him noticed by the management and he was quickly promoted as a result of his care and attention.

FIRST IMPRESSIONS MATTER

Now, here's the thing about what you look like. We all know that beauty comes from within. But we also know that people are people and they judge on appearances first. So although I do agree with people who say it shouldn't matter what you look like, the reality in the workplace is that it does. I truly believe it's important to present yourself well. Look like you care and have respect for yourself. Look the part. Find out how you're expected to dress. If a neat appearance is required, then make sure you take the necessary steps. Don't ever think it doesn't matter. It does, and a business will always have a dress code, even if it's informal. Be prepared and take pride in your appearance.

IT STARTS WITH YOUR HAIR

Your hair says a lot about you and, unless you are an artist or rock star, having a decent haircut is usually required to look professional. Remember, it's not always just what you think; sometimes it's what others think that matters.

A good haircut can enhance your confidence and self-esteem. I advise you to try to match your hairstyle to your lifestyle. Think about your respective career. Distinctive hairstyles have helped several celebrities to gain fame and notoriety, but they aren't that common in the professional world. It's important to keep an eye on what will be deemed suitable for the type of environment you will be working in. Also think about the time needed to style your hair and the maintenance it needs. During working hours you should be focusing on your job, not your hairstyle.

WASHING YOURSELF

It's an unfortunate truth that some people don't always take a shower or use deodorant. But it's just inescapable that in any role, front or back of house, you need to be clean and tidy. Most jobs will require you to work closely with colleagues and/or customers. Be respectful in this area, as you would like others to be with you. Unpleasant smells are very off-putting and can affect other people's performance or discourage customers, which in turn can negatively affect the business. Give yourself time to have a shower and use shower gel, shampoo and conditioner. Dry yourself properly and use deodorant each and every time.

SHAVING

Now to the sometimes controversial subject of shaving. For me, shaving is ultra-important, but many people think that having stubble is cool and part of their creativity, style and character.

If you work in a service industry, being well groomed is important, as it suggests that you are clean. In particular, if you work in the food or medical industries, being unshaven may put people off using your services. If you want to have a beard, that's perfectly fine. But making sure it's evenly trimmed and defined around the neck means you have taken care of yourself, which is a good signal of pride.

MAKE-UP, PERFUME AND AFTERSHAVE

You're changing your life, and being professional is important, so don't overdo the make-up. It distracts colleagues and superiors from your professional attributes. Stick to fuss-free work make-up.

Don't use false eyelashes; they may fall out and embarrass you or, worse, end up somewhere you don't want them.

Look professional, not seductive. Don't wear heavy make-up or perfume. If people notice your make-up, perfume or aftershave, it's probably because you're wearing too much.

POLISH YOUR SHOES

Rather than giving you some detailed guidance on how to polish your shoes, it's simple: polish on, leave for five minutes and polish off.

More to the point is why you should polish them. Of course, it's about preserving the shoes by feeding the leather, thus saving you money in the long run, and protecting them also helps to keep the shoes and your feet dry. But most importantly, they are an extension of you. If you can't be bothered to keep your shoes clean, then what else do you not bother to clean? It's a simple thing to do but very often missed, and it's not good enough. Looking after your feet is highly important, as you can be on them for a long time during your working day.

SOCKS

Good socks will help you to stay comfortable and keep your feet dry by drawing away moisture. Keep away from 100 per cent cotton socks, as they retain moisture and can cause pain and discomfort. A good blend of cotton, wool and silk with synthetic materials seems to be the way forward. Even if your budget is tight, save up and buy some decent socks that help your feet to breathe and are comfortable. Changing them daily and making sure you have clean feet will also help you to stay fresher.

WASHING AND DRYING LAUNDRY

Washing your clothes seems easy, but many people don't place any importance on it. Looking fresh, clean and healthy all adds to your chances of success.

You need to look clean and professional when you arrive for work even if you get changed for duty. Your colleagues, line managers and customers might see you arrive and therefore their opinion of you will start to form before you put your uniform on.

Washing your clothes correctly makes them last longer, the colours won't fade and the shape will be retained.

CARE LABELS AND PRODUCTS

All clothes have care labels with helpful information that can help you save time and money, so here's a quick guide for you.

Clothes that can be bleached are easier to clean, such as whites. If the label doesn't mention bleach, it means it will be safe to use when needed. If a label says use non-chlorine bleach, when needed you will need to buy a safer version of bleach. Always follow the directions on the bleach or packet of stain-remover, as sometimes clothes labels will not give any details.

I would recommend using a decent washing liquid and fabric conditioner. Not too expensive and not too cheap. You will find something that suits your needs and budget, and that works for you. Sometimes the value range doesn't do the job. Look for deals; there are often 2-for-1 offers, and you know you will need the products again, so take advantage of them.

HOW TO WASH AND DRY YOUR CLOTHES

Separate your clothes into dark colours, whites and light colours.

With the door open, select a temperature setting based on the colour and type of clothes you are washing. Most clothes will have a label that lets you know the temperature to use. If you are unsure, then put them aside and wash in a colder cycle. It's not worth risking your clothes; this particularly applies to woollen jumpers.

Here are the three settings I generally use:

Whites – 60°C on the whites setting.

Darks – 30/40°C on the colours setting.

Woollens – 30°C on the woollen setting. (It automatically uses a different spin setting, which helps stop shrinkage.)

Put your clothes in, making sure the setting you have selected corresponds to the clothes you're loading. (Tip: don't overfill the machine. You'll know it's too full if you have to ram the clothes in. Your clothes won't get properly clean and it may break your machine.)

Add your detergent now if you are filling a plastic device (ball) that you have to pop in with the clothes. If not, pour it into the drawer above. Close the door.

Add the conditioner. Now, a lot of people don't use this, but I would recommend it as it softens your clothes, and makes them last longer and smell nice. Again, clean-smelling clothes are important.

Switch the machine on and make a note of the time it finishes. Try not to leave the washing in once the machine has finished. It's easier to dry with the least amount of creases if it comes out quickly.

HAND WASHING

Delicate clothes sometimes require a hand wash. It's simple. Using warm water and a little detergent, soak the garments a little and then wash them gently. Remember to rinse them well, making sure there's no soap left. Then, using warm water again, add a little fabric conditioner, soak again and then rinse well as before. Squeeze out and shake. If hand washing is required, then using a tumble dryer is not recommended; dry by hanging up as directed below.

DRYING

Once the machine has finished its cycle, take each item of clothing out and give it a good shake. It's important to help get any creases out, as it helps the clothes to dry and will reduce ironing. If you're using a tumble dryer, put the clothes into the machine, select a low heat for delicate clothes or medium for most other garments and turn on the machine. Keep an eye on the clothes, checking from time to time to see if they are dry.

If you are not using a tumble dryer, hang as much as possible on hangers or on a clotheshorse. Hangers help dresses, blouses, shirts and T-shirts keep their shape. Clotheshorses are good for trousers, underwear, socks, etc.

FOLD AND/ OR IRON

Once your clothes are dry, try to fold and put them away or iron and hang them away as soon as possible. It gets the job finished and also means they are ready to use and you have cleared the space around you.

HOW TO IRON A SHIRT OR BLOUSE

Start with a clean shirt or blouse. When the garment comes out of the washing machine or dryer, shake it out, smooth it with your hands and hang it on a hanger. Do the top button up to keep its shape.

Fill your iron with water. (Tip: bottled water sustains the life of your iron as it contains less minerals, which can build up and cause blockages and spitting.)

Allow your iron to reach the appropriate temperature, check your garment and the settings on the iron. Remember you can always turn the iron up.

Have some hangers handy so you can hang the garment up as soon as it's finished.

It's handy to have some starch to spray on but not compulsory. Spray on the shirt or blouse before removing from the hanger.

Lay the collar flat on the board and run the iron from the tip of the collar to the middle on both sides.

Press the back of the shirt, where your shoulder blades would be.

Then, take each sleeve in turn and place on the ironing board and press each side, making sure the crease runs down the centre of each arm.

Press the cuffs of the garment if it has them.

Position the body of the shirt on the ironing board, starting with the buttonhole panel first. Run the iron along from the bottom upward towards the collar. Try not to pucker along the buttons. You may find it easier to turn the buttons upside down in order to do these smoothly.

Move the shirt position along to the back panel and press as before.

Move the shirt to the last panel and press.

Return the shirt to the hanger and do up the top three buttons.

GETTING STARTED

DON'T PANIC IF THIS ALL SEEMS A BIT

OVER
WHEL
MING

GETTING STARTED IN YOUR NEW JOB

Now that you've had a chance to think about the responsibility you've agreed to take on and what is meant by being professional, it's time to get started!

WHAT YOU NEED TO TAKE WITH YOU

You need to know whether you should take anything with you. I would suggest a notepad and pencil are essential, but uniform, paperwork, passwords, money for lunchtime and your locker key are other possibilities that spring to mind.

THE BASICS

You need to find out the basics of the job: where things are, where things should be, how the company likes things to be done. There are different sources of information, such as a company handbook, documents with procedures and policies, terms and conditions. Ask what the basics are and get down to learning them.

TERMINOLOGY AND EQUIPMENT

With most jobs, there will be terminology and certain pieces of equipment that you might be unfamiliar with. It will pay dividends to get to know these as fast as you can. For example, if you were becoming a barista, you would need to learn the names of all the coffee varieties, such as flat white, latte, cappuccino, macchiato, espresso, etc. Then there's the coffee machine; I'm sure you'd need to learn all its parts for cleaning purposes. Doing your research ahead of time might help you get the job, and certainly before you start it would be a good idea, if possible, to visit another establishment to ask some questions.

COMPANY INDUCTION

But don't panic if this all seems a bit overwhelming. In order to help you gather all this new information, the company will often give you an induction course. This usually takes place within the first week of employment. It is their chance to fill you in about all the practical details I've mentioned above and then also tell you a bit more about the business you'll be working for: what they stand for, what they want to achieve and how they believe they will achieve it. Sometimes they will show you around the company and demonstrate, perhaps with a short film, what's been achieved so far and explain different aspects of the organisation.

The induction will be used to spell out what's really important to the business, such as:

WHY DOES THE COMPANY EXIST?

Usually, the answer is to deliver a profit to the shareholders through the management's ability to grow its earnings (in other words, its profits and share value). You might not think you're directly involved, but I'm here to tell you that every employee has the ability, and responsibility, to influence the bottom line. You should apply yourself to doing so, positively, every day.

MISSION STATEMENT

A mission statement spells out the organisation's overall goals, which should in turn influence all the actions of the organisation, providing a guide to decision making. It informs employees and customers about the company's belief system and provides a context for strategy formation and implementation.

TARGET MARKET

A target market is a group of customers towards whom the business has decided to aim its services or products.

COMPANY SERVICE OR PRODUCT LINE

The company will explain what they offer to their customers and how they deliver the service to them. It is vital that you understand this so you can provide the service at the appropriate level or follow the production principles. Ultimately, this is what it's all about: delivering a service or product to the customer and making a profit in doing so.

LEGAL ISSUES AND COMPLIANCE

This means anything that by law the company and its employees need to be aware of, abide by and practise on a daily basis, including, for example, health and safety regulations.

HEALTH AND SAFETY REGULATIONS

A list of rules and regulations will have been drawn up with the aim of keeping all employees and customers safe and free from risk. Remember, attention to the health and safety policy is not just about obeying the company's rules; it is the law. But it also makes good business sense.

By applying the rules and regulations you will:

- REDUCE ACCIDENTS

- REDUCE ABSENCE FROM WORK

- IMPROVE PROFIT

- BE MORE EFFICIENT

- IMPROVE THE BUSINESS PROFILE TO CUSTOMERS, CLIENTS, INSURERS AND ENFORCERS

You will be asked to learn the health and safety management system. Make sure that you do so, as it may be that you lack experience and trained judgement, and need some good advice, information and supervision.

OTHER AREAS THAT MAY BE COVERED BY YOUR INDUCTION PROGRAMME

Introduction to terms and conditions – this will include your salary, holiday entitlement and company policies regarding bereavement, sickness, etc.

Emergency contact details – you'll have to provide these in case you are involved in an accident or are taken ill

Completion of government requirements such as providing your P45 or P60

Set up of your payroll details – for this you will need to bring your bank account details

Introduction to key members of staff

Specific job role training

HIERARCHY

In most workplaces there will be a system of hierarchy. This means that members of staff are ranked according to their status and authority. In others words, some people are higher up, paid more and can tell others what to do. They are classed as line managers, bosses, etc.

In my opinion, hierarchies are necessary so that various people at different levels take responsibility. If such systems weren't in place, there could be anarchy and chaos, and nothing would be achieved. It could also lead to a dangerous workplace. Can you imagine being in the army and not having a hierarchy? There would be bullets flying around everywhere. Line management is essential in every workplace; people taking different levels of responsibility at every stage is vitally important for safety and for work to flow smoothly.

You shouldn't fear or disrespect hierarchy. It's a good thing. It is not compulsory to like your boss or line manager, but hopefully you will find some common ground and respect them. You need to speak to them with respect, and they should do the same in return. Just because they are your line manager doesn't give them the right to be rude to you. You should always be able to go to them and discuss issues that arise. But think about when you do this and make sure it's an appropriate time.

UNDERSTAND THE RHYTHM OF THE WORKPLACE

Each place of work and job is different, but understanding the rhythm of the workplace is important. You need to learn to anticipate the busy times and respond when there are deadlines to be met. Make sure you apply yourself appropriately when you are needed. Talking and messing around when you should be concentrating on a task can be very frustrating for your colleagues and employer, and it demonstrates a lack of focus and commitment. It also shows a lack of understanding of the role and what needs to be achieved.

ROUTINE

Establishing a routine is really helpful; understanding what needs to be done and in which order is a bit like time management. Doing something routinely means doing it regularly. Also, there might be a set of procedures to follow at work. This would be a routine. Getting into a routine can build confidence and increase your output or productivity.

GETTING USED TO SHIFT WORK

Shift work is a common employment practice designed to make the most of a 24-hour period. It might be that you have to work different shift patterns. This takes some getting used to, particularly if you haven't worked for a while.

Shift working needs to be considered carefully if you have health issues, and it can put a strain on marital, family and personal relationships.

That said, there are also many positives to shift work. For example, working in the afternoon and evening leaves the morning free to do chores, pay bills, etc. You can also have hobbies that work well with a shift pattern of working.

The point is to think carefully about your life and what suits you best. I personally like shifts because I like my days to be different, and therefore they work for me.

The secret is getting your rota in advance and making sure you remember what shift you're on. There is nothing more frustrating than turning up for work when you're not supposed to.

LEARNING WHILE ON THE JOB

In most jobs there's something to learn. Don't be afraid of this. Try to be excited that you're expanding your capabilities. Be prepared for task-related learning. There will usually be some instructional reading, particularly if you're doing an NVQ-based apprenticeship. The more effort you put in, the more you'll get out of it.

If you feel as though you aren't making enough progress, it can be disheartening. You must get over that, though, and try to ask questions, take notes and demonstrate how you are improving. If you feel you're not making progress, don't wait to be pulled up about it or told off. Ask a colleague or your superior for help or guidance. They will respect you for it.

TAKE NOTES

Taking notes is so important and not often done in the workplace. When learning, you can only retain so much knowledge at a time. Taking notes not only helps you remember what you've learned but also demonstrates that you're taking the role seriously. You can review your notes at a later stage or the next time the task is required; they will remind you about what you need to do. There is a section at the back of this book that you can use for writing notes.

HOMEWORK

Work-based homework is not like school homework. It is usually far more interesting because it's related to something you're passionate about. Do it with enthusiasm and care. It will help you to perform better and with more understanding. You may or may not be being assessed on the homework, but you will be expected to learn whatever task you've been given. Again, put the effort in, as there will be certain expectations of you.

TRAINING PLAN

Many people would laugh at this. Usually a training plan is provided for you. I am suggesting that you write your own. Having a training plan that you own is vital. It's worth sitting down with someone and setting out your own personal goals as well as your work-based goals. Then plan the best way to achieve them (adding in some progression and potential challenges along the way). You may need to identify some additional training that is required to overcome the challenges and achieve your goals.

TEAM WORK AND BEING A TEAM PLAYER

Being a good team member is vital for the success of the business or organisation you work in. Teams can be anything from the group who work in the same part of the building, to the many different people across the company (perhaps some you've never even met) who do the same job in different areas.

A team needs to:

- PULL TOGETHER
- BE CONSISTENT
- BE CLEAN AND TIDY
- COMPLY WITH COMPANY RULES

Come to work ahead of schedule, positive and ready to contribute fully to the team's objectives and high performance. This means you have a greater chance of being successful both within your team and as an individual.

LISTEN

MAKE SURE YOU
LISTEN CAREFULLY
TO ALL INSTRUCTIONS.
IT'S EASY TO MISS
THE POINT ENTIRELY
IF YOU AREN'T
FOCUSED.

QUESTION/CHALLENGE

IF SOMETHING'S UNCLEAR TO YOU, IT'S FINE
TO QUESTION OR CHALLENGE, BUT TRY TO
BUILD ON AN IDEA RATHER THAN KNOCK
SOMEONE DOWN.

ASSIST

WHENEVER POSSIBLE BE HELPFUL —
YOU NEVER KNOW WHEN YOU'LL
NEED HELP.

GET INVOLVED

WHEN DISCUSSIONS ARE TAKING
PLACE, DON'T SIT ON THE
SIDELINES; GET INVOLVED WITH
YOUR TEAM MEMBERS AND HAVE AN
OPINION.

GET STUCK IN

WORK HARD AT ALL TIMES, DON'T BE SHY
AND DON'T BE TOO PARTICULAR WHEN
TAKING JOBS ON.

RESPECT

RESPECT EACH MEMBER OF YOUR TEAM. DON'T FORGET, IT'S NOT COMPULSORY TO LIKE EVERYONE, BUT RESPECT SHOULD ALWAYS BE GIVEN. GIVE EVERYONE SPACE TO TALK AND GET HIS OR HER FEELINGS, IDEAS OR OPINIONS OUT.

BE HONEST

ALWAYS BE HONEST, PARTICULARLY BY OWNING UP TO YOUR MISTAKES. EVERYONE WILL RESPECT YOU FOR IT.

CELEBRATE

WHEN SOMETHING IS ACHIEVED, CELEBRATE WITH YOUR COLLEAGUES. IT'S GOOD FOR MORALE AND STRENGTHENING YOUR RELATIONSHIPS. JUST DON'T OVERDO IT!

GOLDEN RULE

YOU'RE ONLY AS GOOD AS YOUR WEAKEST LINK — DON'T BE THE WEAK LINK!

COLLEAGUE BUDDY

Try to find a colleague buddy. This isn't about friendship but about professional support. I have always found it helpful to have someone in the workplace to go to if things aren't going particularly well.

They will be able to offer you:

- SUPPORT
- ADVICE
- AN ALTERNATIVE OPINION
- ENCOURAGEMENT
- MOTIVATION

So how do you go about finding a buddy? Well, it's not always commonplace to have a buddy scheme, so tread carefully. See if there is such a scheme first and, if so, join it. If there isn't such a company policy, then observe your colleagues. Work out who has the skills you're looking for, such as patience and experience, and someone with a suitable personality.

Once you've selected someone, be clear about your request. What would you like them to do, when and for how long? Most people will be happy to oblige.

GETTING FRIENDS AND FAMILY INVOLVED IN YOUR PROGRESS

I like my friends and family to be involved with my progress. Right from the start, if they know what I'm doing, they will try to help, even if it's just by giving me a pat on the back from time to time. It might be that you don't see your family or you have fallen out and are not speaking. It could still be a good idea to write to them. Even if you don't wish them to be involved, you're making a positive change and it's good to let people know how you are moving on. You never know how it might change people's perceptions of you and can help to build bridges, if that is something you would like to happen.

COMMUNICATION

Good communication is essential in life. Learning to express yourself is so important and is an ability that needs to be nurtured and improved all the time. If care is not taken, it's easy to be too short, frank, unprofessional and not pleasant to be around. If you're not careful, bad communication can be detrimental to your position within the workplace.

Whatever the communication tool you choose, stay in touch – communicate with your employers and colleagues, with your friends and family. We have developed communication skills that are way in advance of any other species. Used wisely, they can be incredibly helpful. Used badly, they can be very damaging.

METHODS OF COMMUNICATION
FACE-TO-FACE CONVERSATION

The ability to have clear and constructive conversations is an invaluable skill in the workplace. You need to be able to discuss issues and concerns, talk through expectations, brief work to be done and compliment people once it has been completed. Poor verbal skills can lead to conversations that undermine confidence, determination and progress. Below I give some examples of the types of conversation that can occur in the workplace.

FUNCTIONAL

A functional conversation is one that takes place because something needs to happen. These are usually work related and are sometimes documented. Always take a functional conversation seriously; take notes if necessary and listen carefully.

BRAVE

A conversation that is difficult but necessary. There are different examples of this in the workplace and they could include having to talk to or with a colleague or boss about personal-hygiene issues, poor performance or a lack of understanding. It could also mean negotiating a very difficult situation such as an emergency.

ANGRY

Anger can play a big part in conversation, but it can be destructive. Try to think through a situation from all angles so you can keep calm and understand other people's points of view.

BETWEEN STRANGERS

Sometimes an awkward conversation can bring good opportunities. Be calm, friendly and polite. Once your nerves have gone it can often be a positive experience for all parties.

BANTER

Banter is informal, non-serious conversation that is usually fun and delivered in a light-hearted manner. Be careful not to alienate people, make fun of them or hurt anyone's feelings. Banter should be inclusive, not exclusive.

WARNING!
WATCH OUT FOR NARCISSISM

Narcissism is a common failing and I will hold my hand up to being guilty of this at times. It means turning a conversation so it's all about me, closing down questions and not wanting to hear other people's opinions. It can be seen as rude, selfish and arrogant. This isn't nice, so try to keep it open at all times.

CONVERSATIONS WITH YOURSELF

Yes, that's right: sometimes you have conversations with yourself. No, you're not mad; you have them to help yourself solve an issue or fill the silence around you. It's perfectly normal and healthy. If you start having them with the carrots in your fridge, then you should be worried!

TELEPHONE CALL OR TEXTING

Being able to use the phone appropriately is hugely important, particularly if you are running late or don't feel well. A short call or a text to explain the situation well ahead of time can reduce inconvenience for your employer by enabling the company to change rotas, schedules, etc.

Not phoning or texting is unforgivable unless you absolutely cannot do it. But generally there is no excuse, and such a failure will result in a black mark against your name. If it happens too many times, you will be disciplined.

EMAIL

Emails are the modern equivalent of the letter. But they are instant and non-retrievable, so time should be taken when writing an email. Don't send it straight away; read it through to make sure you've been clear and concise.

FRUSTRATION, DISAPPOINTMENT AND ANGER

We all get frustrated from time to time, sometimes with ourselves. Maybe you're unhappy with how you performed at work, maybe you've disappointed yourself for some reason or are disappointed in someone else. There are lots of reasons why frustration and disappointment can creep into the mind.

How you deal with these emotions is vitally important. You can't easily conceal frustration or disappointment. They can eat you up inside and then manifest themselves in anger or an outburst of raw emotion. Please be careful, as this usually creates a lot of tension, doesn't necessarily come out the way you plan, almost certainly damages the relationship with the person you direct your frustration at and makes matters far worse.

Count to ten; think about what you're going to say, or, even better, write it down. Discuss it with someone else and they might help you think it through and see the situation differently.

Never threaten people. Always think about your body language and calm yourself down as it's all too easy to make the situation worse. Anger in the workplace is a sackable offence. It would be a massive shame to be the one to lose your job because you couldn't control your temper.

RESPONDING POSITIVELY TO FEEDBACK

If you respond positively to feedback, even negative feedback, your employer will see that you're serious about your job and keen to make progress. Show them that you're able to take their comments on board and make the appropriate changes. Don't be defensive or allow it to get you down. See it as another opportunity to learn.

All in all, if you come to work with a good positive attitude, you should enjoy what you're there to do, people will respect you and you should achieve success.

FINAL MESSAGE

Well, congratulations and goodbye, reader.

It's been fun, hasn't it? Well, maybe not fun, but I have learned a great deal while writing this book. Taking my own advice should be interesting, but I do believe that I will try to walk my talk for sure.

We dipped into some pretty intense places and you will have uncovered a few home truths along the way, including some really positive ones, while identifying some things to work on. One thing is for sure: you are amazing, even though when you started reading this book you may well have felt a bit lost or unfocused, certainly not confident in your opportunity to find and keep employment. I hope you feel much more positive and able to get yourself to a place where you can find a job and keep it.

To measure how far you may have come maybe you could ask yourself a question.

After reading this book – how do you feel?

A No difference – won't be using the book at all

B A bit better – thinking it all over and might start using some of the thinking

C More positive and wanting to make changes and use some of the advice in the book

D Amazing and I am using the book to move myself forward and find employment

If you answered **A** I hope that you could revisit the book when you feel you are ready to. If you answered B to D I am very pleased for you. I really would love to know how you get on. Keep up the good work and I am sure you will succeed and be in full-time meaningful employment very soon.

I wrote this book so that you will learn from some of the things I have picked up over the years. Of course, everyone has their own experience and there's more than one way of doing things.

The exercises all have a purpose, and it would be good to reflect on the ones that helped you the most. If you found some particularly hard it might be worth revisiting them and trying them again. If there are sections that you haven't tackled yet, don't worry. Just go back and give them a go. Refer back to the personal projection plan: it will help remind you why you started this and it will encourage you to keep at it.

You should share your thoughts, notes and templates with someone whom you feel comfortable to discuss them with. They could help you build a stronger story, CV and covering letter. In addition they will challenge or help build on what you have already achieved on your own.

I hope this book helps you clear your pathway to a better, less stressful life. A life full of opportunity, positivity and happiness. It won't always be easy; life's not like that. There will be times where you feel like giving up, someone at work bothers you or things don't pan out the way you want them to. Remember it's a journey, and sometimes we have to take a side step every now and then. But if you can try and find something you love doing it will feel much easier and more enjoyable. That must be what we all want for our families, our friends and for ourselves.

One thing's for sure: you must not let life pass you by. Live every moment by moment, and learn from everything you do. You can choose to have a working life that is colourful, worthwhile, memorable and meaningful.

Good luck with everything.

SIMON BOYLE

NOTES

ACKNOWLEDGEMENTS

This book would not have happened and been able to help so many people without the dedicated support of the following people and organisations.

Carey Smith

Lydia Good

Nicola Hill

Doug Young

Alison O'Toole

Clarissa Pabi

Ailsa Bathgate

Penguin Random House

PwC

Beyond Food CIC

Brigade Bar & Bistro

DWP

De Vere Venues

Simon Boyle

Simon Boyle is the UK's leading chef that you've probably never heard of.

Trained in the kitchens of the Savoy Hotel, London, and by the culinary legend, Anton Mosimann, Simon has worked across the world for the likes of Unilever as a retail products innovator, bringing to life new products and concepts for world famous brands. Simon has travelled the world with P&O Cruises and has even worked for a Prince in Saudi Arabia.

Simon founded the highly acclaimed Beyond Food Community Interest Company (CIC). Beyond Food has its own innovative social enterprise restaurant in London, Brigade.

Throughout this book, Simon uses his years of experience to help the reader's endeavours to find a real purpose and happiness in life, and a long-term sustainable career to boot.

Brigade

Brigade is a social enterprise restaurant housed in a unique social enterprise hub in The Fire Station on Tooley Street, London, SE1.

First, it is a British inspired Bistro serving delicious, seasonal and sustainably sourced dishes. Second, it is a training kitchen that provides apprenticeships to help the homeless develop skills and motivation to find employment.

Brigade is part of Beyond Food Community Interest Company (CIC), a unique collaboration between PwC, De Vere Venues, Beyond Food and Big Issue Invest. Beyond Food CIC inspires people who are at risk of, or have experienced, homelessness, to gain meaningful employment. Brigade has trained thousands of people and won countless awards.

Beyond Food runs a series of training programmes, which culminate in six-month apprenticeships in the Brigade kitchen followed by seven months in another professional kitchen in London. At the end of the 13 months, our apprentices attain an NVQ level 2 diploma in professional cooking and we help them to find long term employment in the hospitality industry.

Since opening in 2011, Beyond Food has helped countless vulnerable people, some of whom were sleeping rough when they joined the programme. Most were living in hostels and a large number were ex-offenders or had issues with substance abuse. We're passionate about inspiring people to a life beyond homelessness.

www.thebrigade.co.uk
www.beyondfood.org.uk

LAST TRAIN
to
PARADISE

Henry Flagler and the
Spectacular Rise and Fall of
the Railroad That Crossed an Ocean

LES STANDIFORD

THREE RIVERS PRESS
NEW YORK

Published by Three Rivers Press, New York, New York.
Member of the Crown Publishing Group, a division of Random House, Inc.
www.randomhouse.com

Three Rivers Press and the Tugboat design are registered trademarks
of Random House, Inc.

Originally published in hardcover by Crown Publishers,
a division of Random House, Inc., in 2002.

Printed in the United States of America

DESIGN BY LEONARD W. HENDERSON

Library of Congress Cataloging-in-Publication Data

Standiford, Les.
Last train to Paradise : Henry Flagler and the spectacular rise and fall of the
railroad that crossed an ocean / Les Standiford.
Includes bibliographical references.
1. Railroads—Florida—History. 2. Flagler, Henry Morrison, 1830–1913.
3. Florida East Coast Railway—History. I. Flagler, Henry Morrison,
1830–1913. II. Title.

TF24.F6 S73 2002
385'.09759'41—dc21
2001047565

ISBN 1-4000-4947-4

10 9 8 7 6 5 4 3 2 1

First Paperback Edition

This book is dedicated to the memory of the hundreds of "vets" who lost their lives during the Labor Day hurricane of 1935, as well as to the hundreds of Keys residents—men, women, and children—who also suffered and died that day.

⊠ ⊠ ⊠

I would like to express special thanks to Bernard Russell for sharing his memories of the '35 storm so candidly; to Scott Waxman, who envisioned what this book could be; to Robert Mecoy and Emily Loose, who believed in it from the beginning; and to Kimberly, Jeremy, Hannah, and Zander for their boundless support and patience.

Special thanks is also due a number of individuals who lent invaluable support and assistance to this enterprise, most prominent among them: John Blades, director of the Henry Morrison Flagler Museum in Palm Beach, Dr. Robert Gold, Catherine O'Neal, and, especially, Rebecca Callahan.

I am also indebted to Steven Leveen and Dr. Bill Beestig for their close reading and fact-checking of this manuscript.

And the usual inexpressible gratitude goes to James W. Hall and Rhoda Kurzweil, who looked over my shoulder all the way along, correcting, suggesting, and, as always, reminding me of the little blue engine that could.

Contents

Author's Note

THE GREAT TEXAS WRITER and historian J. Frank Dobie once lamented that most historical research amounts to the carrying of the same old bones from one grave to another. And as I have gone about the research and the writing of this book, I have been mindful of Dobie's words every working day. The fact that able explorers have scouted the most promising grave sites in advance can be helpful, of course; but if all the artifacts have been unearthed and duly noted, then it might seem there is little useful work left to do.

Indeed, others have written well and truly of Henry Morrison Flagler and his remarkable life. His part in the creation of Standard Oil has been ably documented, even if many remain unaware that he and John D. Rockefeller were once coequals in that process. And other writers have credited Flagler— through the building of the Florida East Coast Railway and the founding of most of the state's best-known cities, including Miami—with the veritable creation of Florida as we know it today.

Even the story told here, of the building of the Key West Extension to Key West, "the railroad that crossed the ocean," is not an unknown one, at least in general outline, to Floridians who have lived in the shadows of that great undertaking, known someone who helped in its building, driven by car along the route, or read one or another of the many monographs and newspaper articles and personal accounts connected to it.

But if we are talking about the disinterment of old bones here, then this telling of the Key West Extension's birth, and its demise, aspires at the very least to lay them out in a common grave. And if doing so enhances anyone's appreciation of the magnitude of Flagler's mammoth, star-crossed undertaking, then it has been worth it.

By anyone's estimation, Henry Morrison Flagler accomplished a number of remarkable things in his eighty-three years, but it is the thesis here that nothing he did, not even the creation of Standard Oil—the ultimate symbol of corporate wealth and power—can compare to the laying of that last stretch of Florida East Coast Railway track.

Building Standard Oil required great business savvy, a certain measure of ruthlessness, and not a small amount of luck. Building a railroad across an ocean required extraordinary vision, effort, perseverance, and sacrifice. The former qualities, which Flagler used in abundance during the first half of his life, seemed far less central to the second half. Certainly whatever luck visited the undertaking of the railroad across the ocean was bound to be bad.

No one today would undertake what Flagler did, not in this bottom-line world. Nor would they undertake Chartres or Notre Dame, for that matter. Dreaming up a railroad to Key West is the stuff of another era, and its undertaking is the work of another kind of man.

In the impossibleness of what was once called "Flagler's Folly" is also its magnificence. In its final undoing is the significance of tragedy.

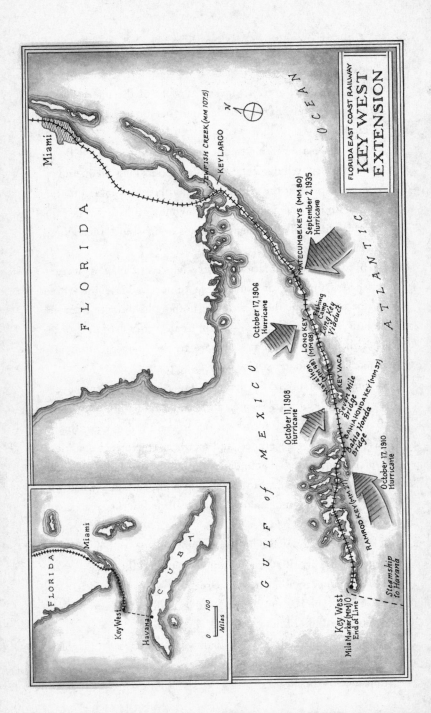

And on the pedestal, these words appear:
"My name is Ozymandias, King of Kings,
Look on my Works, ye Mighty, and despair!"
Nothing beside remains. Round the decay
Of that colossal Wreck, boundless and bare
The lone and level sands stretch far away.
—Percy Bysshe Shelley

Ah, but a man's reach should exceed his grasp,
or what's a heaven for?
—Robert Browning

End of the Line

KEY WEST
Labor Day Weekend, 1935

At about four o'clock in the afternoon on Labor Day Saturday in 1935, Ernest Hemingway, by then one of Key West's most notable residents, thought it time to knock off work on weaving together what an editor had called "those Harry Morgan stories," an undertaking that would eventually be published as a novel titled *To Have and Have Not.* He left his studio, went into the kitchen with its high, built-to-Papa cabinet tops, to pour himself a drink, then walked out onto the spacious porch of the two-story home on Whitehead Street that he and his second wife, Pauline, had bought in 1931.

The day's work had been good. Now he intended to wind down and have a look at the evening paper.

The weather was typical for late summer in Key West: the temperature in the high eighties, the humidity about the same, but the skies were clear, and there was a sea breeze sweeping over the mile-wide island to soften the heat, especially in the shade of a broad front porch.

It was a newfound pleasure for Hemingway to indulge himself in such a simple fashion, even in his own home. The year before, a zealous Federal Emergency Relief Act Administration official had published a pamphlet intended to boost tourism, listing Hemingway's home among the top twenty-five attractions on the island of some twelve thousand souls.

Though Hemingway well understood the value of cultivating a certain mystique, it had nonetheless galled him to find himself, on the way to or from his workroom on the second floor of a then-unattached outbuilding, staring back at a queue of gawking visitors on the other side of the chain-link fence that protected his property. Thus, only a few days before, and after much wrangling with a city bureaucracy that considered it an eyesore, work had been completed on a stone wall that now marched about the three open sides of the house's corner lot, giving him some measure of privacy at last.

It is easy to imagine Hemingway in a reasonably affable mood that afternoon. "Now that I've gone private," he'd remarked to his longtime handyman, Toby Bruce, once the wall was up, "they might even take me off the tourist list."

And because it was the off-season, there would be no crowds in Sloppy Joe's Bar to annoy him during his late-night rounds. Nor had the "mob"—as he sometimes referred to the annual coterie of friends and hangers-on from the North— arrived to lure him from his work to go on fishing expeditions out to the nearby Gulf Stream or Dry Tortugas, or to an endless round of parties there on land.

Earlier that summer he had turned in a completed manuscript of *The Green Hills of Africa*, which he privately considered his best writing since *Death in the Afternoon*. With publication scheduled in October, Hemingway was eager to see if the public's approbation matched his own. Though he'd had similar hopes for the bullfighting book when it was published in 1932 and had been disappointed by the decidedly mixed opinion of the critics, he was certain he would receive his due this time.

He'd received a nice little bonus in the form of a five-thousand-dollar sale to *Scribner's* for the magazine serialization of *Death in the Afternoon,* things were going well between him and his second wife, Pauline, and he was intrigued with his current project in *To Have and Have Not,* where he intended to bring fictive life to all the Key West lore and legend that he had accumulated since moving to the island city in 1928.

Not a bad moment, then, not by any stretch of the imagination: the end of a good day's effort, a drink in hand, a breezy porch to lounge upon for a glance at the day's events . . . until everything suddenly changed.

STORM WARNING! was the banner headline Hemingway found in front of him, and, just below, the details of a hurricane feared to be coming Key West's way.

In those days, weather forecasting was primitive, by modern standards. The storm, which had formed off the coast of Africa sometime during the last week of August, had moved across the Atlantic, undetected by the likes of modern-day satellite eyes or storm-chasing converted bomber planes, and now it was zeroing in on the United States.

Ships steaming southward to Havana were the first to encounter the disturbance, then a minimal hurricane with winds hovering in the seventy-five-miles-per-hour range. The reports were forwarded by telegraph back to Miami, where, in good time, newspapers had passed along the news. Though there were no computer tracking models to consult, in the Keys the average landmass lay lower than the top of a small child's head above sea level, and any fool—much less Ernest Hemingway—knew enough to get ready for trouble.

The papers reported the location of the storm at press time as just east of Long Island, in the Bahamas, some four hundred miles east of Key West. Hemingway finished his drink, put his paper down, and went into the house to dig out his storm charts, one of which detailed the dates and tracking of the forty hurricanes that had, since 1900, approached Florida during the month of September.

Given the reported rate of speed for the current storm (the quaint practice of naming hurricanes was not adopted by the U.S. Weather Bureau until 1953), Hemingway calculated—without the aid of television newsmen or late-breaking advisories—that he had until noon on Labor Day Monday before the worst might hit.

Hemingway's first concern was his beloved boat, *Pilar,* a forty-foot powered fishing yacht he'd had built to order in a New York shipyard hardly a year before. His game-fishing forays about the northern Caribbean with Pauline and fellow writer John Dos Passos and Key West barkeep "Sloppy Joe" Russell and famed bullfighter Sidney Franklin and so many others were already the stuff of local legend, and Hemingway was prone to discuss the boat with others in a way that sometimes made casual acquaintances think he was referring to a lover.

As anyone who has tried to secure a boat in the face of an advancing hurricane can attest, however, the process is a tedious and frustrating one, complicated by a steady escalation of panic among other owners, many of whom may not have visited their craft in months. And Hemingway, despite his notoriety, found himself no exception. In a piece he wrote for

The Masses, a left-leaning publication of the day, he shares a vivid picture of what he was up against.

> *Sunday you spend making the boat as safe as you can. When they refuse to haul her out on the ways because there are too many boats ahead, you buy $52 of new heavy hawser and shift her to what seems the safest part of the submarine base and tie her up there.*

With the boat attended to as best he could, Hemingway spent the rest of Sunday evening and the following morning feverishly moving lawn furniture, carrying in plants, and shooing the ever-present horde of cats inside his house, then nailing makeshift wooden shutters over all the windows. By five in the afternoon the storm had not materialized, but the double red and black flags that signified an impending hurricane were snapping over the Key West harbor in a heavy northeast wind. The barometer was falling precipitously, and the streets all over the town resounded with the crack of hammers driving nails into shutters, which nervous owners only hoped would hold.

With nothing more to do at home, Hemingway left Pauline and returned to the Navy yard where he'd tied up *Pilar:*

> *You go down to the boat and wrap the lines with canvas where they will chafe when the surge starts, and believe that she has a good chance to ride it out . . . provided no other boat smashes into you and sinks you. There is a booze boat seized by the Coast Guard tied next to you and you notice her*

stern lines are only tied to ringbolts in the stern, and
you start bellyaching about that. . . .

Hemingway was enough of a sailor to know that lines
attached to a few bolts drilled into the deck of a poorly maintained boat could never withstand the pressure exerted by the
winds of a hurricane, but his complaints had little effect on
an already overburdened staff. The harbormaster simply
shrugged and told him he had permission to sink the rumrunner if she broke free and threatened to ram *Pilar*.

Just how Hemingway was supposed to manage such a feat
in the midst of a hurricane was not made clear, but there was
nothing else to be done at the basin. He gave one last baleful
glance at the precariously tied-off rumrunner, then made his
way back to the house on Whitehead Street, left with the very
worst thing to do as a hurricane approaches: wait.

From the last advisory you figure we will not get it
until midnight, and at ten o'clock you leave the
Weather Bureau and go home to see if you can get
two hours' sleep before it starts, leaving the car in
front of the house because you do not trust the rick-
ety garage, putting the barometer and a flashlight by
the bed for when the electric lights go. At midnight
the wind is howling, the glass is 29.55 and dropping
while you watch it, and rain is coming in sheets. You
dress, find the car drowned out, make your way to
the boat with a flashlight with branches falling and
wires going down. The flashlight shorts in the rain,
and the wind is now coming in heavy gusts from the
northwest. . . . you have to crouch over to make
headway against it. You figure if we get the hurri-

cane . . . you will lose the boat and you never will
have enough money to get another. You feel like hell.

Hemingway's preparations, and his premonitions, were well
founded, as it turns out. On Matecumbe Key, some eighty
miles to the north and east of Key West, the full force of the
storm had already begun to sweep ashore. Residents of Islamorada, the principal settlement on Matecumbe, stared in disbelief as their barometers plunged from a normal 29.92 down
to 26.35 inches of mercury, the lowest reading ever recorded
in U.S. history.

Local residents had endured fearsome tropical winds before,
and had some idea of what to expect. But this was no ordinary
storm, and the situation on Matecumbe at this particular time
was especially dire.

As part of his New Deal, Franklin Roosevelt's Federal Emergency Relief Administration had put 650 indigent World War I
veterans to work on a highway building project near Islamorada, using a route that paralleled what was then the only
link across the low-lying islands between Miami and Key West:
153 oceangoing miles of Florida East Coast Railway track, a
daunting project completed in 1912 by Henry Morrison Flagler, and often referred to as "the Eighth Wonder of the World."

The work camp for the new highway-building project was
situated on Matecumbe Key; offices were nothing more than a
few hastily constructed outbuildings; the workers, most
brought in from Northern cities and lacking any notion of
what horrors a hurricane might bring, were housed in flimsy
tents.

And the storm, as if sensing the most vulnerable place to
come ashore, had drifted slightly east of Key West. As the

monster began to hammer the camps, panic became the order of the day. Winds were already gusting over 155 miles per hour, the contemporary threshold for a Category 5 super-storm. Such categories had not yet been developed in 1935, but as one former resident remarked, "we saw pretty quick this was going to be a son-of-a-bitch."

Power was out and tents were shredding away like tissue, leaving men to cling to staves, mangrove roots, even train rails, to keep from being blown off into the rain-lashed darkness. One vet tied himself to a tree with his belt, but his reprieve was short-lived when the wind tore the tree up by its roots and carried it, and him, away.

Melton Jarrell, one of the camp workmen, described his ordeal in a story carried by the *Miami Herald*: "I made for the railroad and hung onto it. A heavy sea came along and washed it up and as it settled back down, it pinned my left leg under it. In horrible agony, I decided to cut my foot off but I couldn't get to my penknife. After that, I passed out."

⬚   ⬚   ⬚

At nearly the same time that Jarrell had been about to hack off his own foot with his pocketknife, a permanent Islamorada resident, Bernard Russell, then seventeen, had run for shelter in a lime-packing shed, along with his parents, three sisters, and an uncle and his five children. The Russells, whose patriarch, John Henry Russell, had emigrated to the Keys from the Bahamas in the middle of the nineteenth century, were part of an extended farming and fishing clan numbering more than sixty that populated Matecumbe and others of the Middle and Lower Keys.

"We were going to ride it out in the beach house, but my

dad saw how it was going and changed his mind. We headed for the packing house, which was close to the railroad, and on higher ground," Russell recalls.

The notion of "higher ground" is a relative one, given Keys topography. Most of the Keys are not true islands, but mere outcroppings of reef, calcified remains of sea life, which poke just above the level of the sea. Most of the formations are only a few hundred yards or less wide, enhanced here and there by dredging and other means of artificial fill. By most American standards there is little that resembles a rise, let alone a hill, not in the entire 220-mile stretch of the archipelago that runs from Biscayne Bay, off Miami, to the Dry Tortugas, seventy miles beyond Key West. The highest point in the inhabited keys, which range from Key Largo, fifty miles south of Miami, to Key West, another 110 miles south and west, is barely sixteen feet above sea level.

The "higher ground" Russell refers to, then, was that created by the mounding of a few feet of crushed limestone marl and gravel riprap to create the bed for the Florida East Coast Railway, a quarter of a century before the hurricane hit. From that vantage point it was possible to glance north and south and see ocean in either direction.

"We knew when the storm hit—at eight o'clock—because that's when all the clocks and the watches stopped. That's when the water started covering everything," Russell says.

The wind was so strong by that point that the whole packing shed had begun to throb, each pulse growing stronger, like the chamber of a giant heart about to blow itself apart with its next beat. Russell's father ordered the boy to stand against the door, which was vibrating like a tuning fork, ready to spring off its hinges. Russell continues:

This shed was built up on a kind of platform to help with the loading and unloading, but while I was trying to lean into the door to keep it from blowing in, I felt something wet and looked down to see water pouring under the sill around my shoes. I thought I knew what that meant, but I put my hand down in it and tasted it just to be sure. It was salt. That meant the water had risen overtop of the whole island. I told my Dad and he shook his head.

"We'll have to get out," he said. "Or drown."

He told everybody to grab hold of somebody else and not let go, no matter what. I grabbed my sister, who had her little two-year-old boy in her arms. I told her to let me hold the boy, but she shook her head and wouldn't let go. I couldn't argue with her. It wasn't that kind of time. So I held her and him together, as best I could, and out we went.

The noise and the wind were unbelievable. The minute we were outside, the wind took us, and we began to spin around, all three of us. I tried to hold on to them but I couldn't. We were in midair as I watched her being pulled away from me. She still had hold of her little boy. And I was trying to reach her, but I just couldn't. It was like being in a nightmare.

* * *

That railroad embankment where Bernard Russell and his family had run for safety, where the men of the work camps cowered under the force of the storm, constituted the sole lifeline for the thousand or so people stranded on the Matecumbes that night. After all, the workmen had been brought to the Keys to build a highway bridge that would replace the ferryboats linking Islamorada with the mainland. With the bridge unfinished and boat traffic out of the question in such weather, only the railroad remained as an escape route from the tides that now threatened to obliterate the Middle Keys entirely.

In Miami, officials had earlier that day begun to grasp the seriousness of the situation, and in response to frantic pleas from work camp supervisors, the Florida East Coast Railway had finally dispatched a rescue train from the North Miami yards.

Because it was a holiday weekend, however, there was some delay in rounding up a train crew and assembling the necessary equipment. It was nearly 4:30 P.M. before a locomotive pulling six passenger cars, two baggage cars, and three boxcars finally left Miami. Scarcely had the train gotten under way when there was another maddening ten-minute delay at the crossing of the Miami River, where a turntable bridge yawned open to allow the passage of Labor Day pleasure craft below.

By this time the initial storm bands had begun to push ashore as far north and east as Homestead, the jumping-off point at the southern tip of the Florida mainland. As engineer J. J. Haycraft guided Old 447 through the increasingly intense squalls, his misgivings grew, for he was a fourteen-year veteran of the Extension and had seen his share of tropical storms. The otherworldly gray-green cast of the sky before him told Haycraft that this was likely to be the granddaddy of them all.

Given the intensity of what he sensed coming at him, Haycraft reasoned that it only made sense to shift the big locomotive from the front of the train around to the rear. That way he could back his way down the single-line track that crossed the

Keys, and, after he'd piled everyone on board, could pull straight back north, hell-bent for leather, able to use the engine's headlamp to guide the way through the oncoming darkness. It might have been a prudent decision, but going through the switching process in the Homestead yards took another precious fifteen minutes.

By now it was nearing five-thirty. The winds had risen beyond the point of exerting mere physical pressure. The force was such that any matter capable of movement—guy wires, stays, eaves, power lines, trees, and timbers—had begun to vibrate, whine, and moan, each element calling out in its own characteristic voice and pitch. The result was a harmony of dread that anyone who has lived through a Category 4 or 5 storm can never forget—they will tell you that in many ways, that unrelenting, awful sound is the very worst part of all.

"It's like a freight train roaring forever right over your head."

"Like an avalanche that never, ever stops."

"Like your head is ready to blow apart."

"Like hell on wheels, bud. And it's got its eye on you."

Visibility was near zero, the rain blasting through the open cockpit like needles in Haycraft's face and eyes. Even on relatively broad Key Largo, the winds had driven the tides hundreds of feet across the flattened landscape to lap at the verge of the rocky right-of-way. Though he'd had to cut his speed back to twenty miles an hour once he'd left Homestead, Haycraft pressed on, mindful that he was the only hope for those trapped by this ungodly storm.

By the time the train approached Windley Key and the first major span over open water, it was nearly 7:00 P.M., and light was virtually gone, save for the strobelike flashes of greenish lightning. At the approach to Snake Creek Bridge, Haycraft caught sight of a group of refugees crowding up to the track, panicked by the water that now threatened to engulf them.

Though his primary mission was to reach the work camps and the 650 who were counting on him, Haycraft did not hesitate. He brought the big engine to a halt and waited for the refugees to be loaded on board.

Haycraft's instincts might have been understandable, but as certain philosophers have noted, nature is short on understanding. And so are the fates.

When Haycraft fired up Old 447, the train started forward, then there was an ungodly crash and the engine lurched abruptly to a halt as if a giant hand had seized it. Haycraft worked his throttle frantically, staring about in disbelief at his crew. No wind, no matter how strong, could hold back a roaring locomotive's thrust. "I didn't know whether we'd had a wreck, a washout, or what," the engineer said. "We might have been at a bottomless pit, it was so dark."

For a moment the men wondered if they might be in the grip of some force that went beyond reason. Then trainmaster G. R. Branch clambered up from the floor where the impact had thrown him and pulled Haycraft away from the engine's controls to show him what had happened.

A thick gravel-pit boom cable, which normally passed high above the tracks, had sagged when one of its supports blew down in the storm. The cable had somehow cleared the tender car behind them, but had swept into the open cab of the engine where Branch had been standing. Had they been going any faster, the cable might well have cut the trainmaster in half. As it was, Branch had been thrown to the floor by the impact, and as the train had rolled on, the cable had tangled in the superstructure of the cab, eventually dragging the engine to a halt.

When Haycraft realized this, he tried backing up, but the

impact had virtually welded the cable to the engine. It took nearly an hour for crewmen to locate the proper tools to cut the thick cable free.

Less than twenty miles remained between Haycraft and his destination now, but in that short distance, Old 447 was to traverse the great gulf between the known and the never-before-encountered. The full force of the storm had begun to cross the Matecumbes now, the barometer plunging to record lows, crewmen forced to work their jaws against the sudden popping in their ears.

The winds were approaching an ungodly two hundred miles per hour, far beyond any forecaster's expectations. The rain was a horizontal force, as painful as a sandblasting, so much moisture aloft that it was difficult to draw a breath.

A visitor to one of the work camps that day, Charles Van Vechten, recalls seeing the train that was to have rescued him as it passed him by:

> You can't imagine how awful it was. At noon we were told to expect a storm . . . but that a train would arrive in time to take us out. . . . We packed up in the afternoon, and assembled, ready to leave, but the storm hit before the train got there. When it did, I guess it was about 8:00 P.M., and it was pitch black and blowing like fury. I saw bodies with tree stumps smashed through their chests, heads blown off, twisted arms and legs, torn off by flying timber that cut like big knives. When the train came I dug into the sand to keep from being blown away. I saw the sea creep up the railroad elevation like it was climbing a stairway. The train went on past, heading for the other camps on Lower Matecumbe, I guess.

Haycraft never saw Van Vechten, of course. From his perspective, water covered virtually everything. Over tracks that had once stood seven feet above sea level approaching the Islamorada station, breakers were crashing.

By all appearances, Haycraft was now piloting a rocking train—at one to two miles an hour—across the surface of the ocean itself, and even he had begun to despair. How could anyone survive? he wondered. For that matter, how could he?

Then, shouts from his crewmen brought him out of his reverie. As if in a dream, Haycraft caught sight of desperate faces flashing past the engine bays, the hands of men, women, and children clutching toward the train that was passing them by.

Haycraft brought his engine to a halt some 1,500 feet south of the Islamorada station and watched as the crowd stumbled down the rails toward him: women at the front of the pack, as it should be, many of them clutching children by the hand, others pressing infants to their breasts. Something would come of all this effort, then, he thought. Some precious few lives could be saved, after all.

He would load up this band of human cargo and steam northward out of a watery hell, and not let himself think about what wretched others might be clustered on down the line. It was time to cut the losses, get out while the getting was good.

And then he felt the grip of his fireman upon his shoulder, and sensed the panic in the man's shouts. Haycraft turned to see what had possessed the fireman, then caught sight of it out of the corner of his own disbelieving eye. At the same instant, he felt the rumble rising up from beneath his feet, a growling that overwhelmed even that of 447's mighty engine.

A dark wall was rushing toward them, a swath of blackness and evil that seemed to swallow the dim illumination of the locomotive's headlamps. Nearly twenty feet tall it was, and it stretched across the horizon from end to end like the sweep of doom itself.

A tidal wave. The worst that had ever struck American shores. Then and now.

"Lord have mercy," J. J. Haycraft murmured, his hand going instinctively for the throttle. And everything was dark.

The Road to Paradise

The intertwined skein of
design and fate that brought
Ernest Hemingway, Bernard Russell, J. J. Haycraft, the six hundred or more veterans, and scores of others together in the Keys on that ill-fated Labor Day night in 1935 is complex indeed, possible to trace in retrospect, perhaps, but impossible to have foreseen. Rust Hills, literary critic, fiction editor of *Esquire* magazine for much of the last half of the twentieth century, and a seasonal resident of Key West himself, says as much is true of any good story:

"On your way along, it seems there are a myriad of choices cropping up for characters to make, one road after another constantly dividing and offering first this alternate path then that, so from the direction that experience takes, it seems that things could turn out in a million different ways. Once the story is over, though, you can look back, retracing the steps, and see that this ending was inevitable, that every choice along the way led to the one, unavoidable place."

Given history's perspective, then, what happened on September 2, 1935, was not so much an accident as a culmination of forces that had been set in motion many years before. On that fateful day in 1905, however, when Henry Flagler, cofounder of Standard Oil and one of the world's most famous and powerful men, announced that he would extend his far-flung empire by building a railroad across the ocean, few could have anticipated how things would ultimately turn out.

Many immediately dismissed Flagler's intentions as impossible, even lunatic, it is true. But those detractors weren't in the slightest concerned about some storm to come decades later. They were not soothsayers but hardheaded scientists, engineers, and businessmen who thought what Flagler proposed—to build a railroad 153 miles from Miami to Key West, much of it over open water—a crackpot notion on the face of it.

"Flagler's Folly," the press dubbed the project, though the man who proposed it was undeterred. He would press on, though what was to come of his vision—certainly what remains of it today—bears little resemblance to what he originally had in mind. And while he was quick to offer practical justifications for the project at the time, in retrospect it seems apparent that other forces were driving this otherwise sober-sided man.

This much is certain: Most contemporary travelers who drive the 128.4 miles of US 1 that now stitch the Florida Keys, from Homestead at the tip of the U.S. mainland to Key West at the very end of the line, find it one of the most remarkable stretches of highway in the country.

Though the roadway across the Keys may lack the mountainous drama of a few rival Western counterparts—Seventeen Mile Drive, sections of the Pacific Coast Highway—the road through the Keys offers an intimacy with sea and land and sky and a variety of perspective and play of light unparalleled anywhere.

This piece of highway also offers a certain definitiveness, a very inescapable destination: Key West sits literally at the end of the American road, and for most of the twentieth century the "Southernmost City" has sung a siren's song to tourists and travelers, literati and glitterati, grifters and drifters and modern-day pirates alike.

For most travelers, the highway down the Keys is a convenience. For many, the scenic aspect of the drive is a welcome bonus. Certainly, few would give this highway's history much thought.

To tell the average traveler that until the twentieth century the only way down to Key West was by boat would likely elicit a tolerant stare. Adding that once only a railroad ran this very route, where now only asphalt snakes along, might bring a distracted nod from a listener with a fix on Margaritaville.

Yet the story behind the very being of this road may be its most amazing aspect. It is a story that concerns one of the world's richest men, one of the most difficult engineering feats ever conceived, and the most powerful storm ever to strike American shores.

In a sense, the highway is what remains of one of the last great gasps of the era of Manifest Destiny and an undertaking that marked the true closing of the American Frontier. The building of "the railroad across the ocean" was a colossal piece of work, born of the same impulse that made individuals believe that pyramids could be raised, cathedrals erected, and continents tamed. The highway is a ghost really, all that remains of an era when men still lived who believed that with enough will and energy and money, anything could be accomplished.

Quite often, in the wake of terrible natural catastrophe—typhoon, volcanic eruption, earthquake—the press confuses accident with intention and labels a staggering loss of property and life as "tragic." But as Aristotle—were he still around—would be quick to remind us, nature is as immune to tragedy as it is indifferent to man in general.

Tragedy is a human invention, called forth to give meaning to certain catastrophes, those in which unfortunate events are exacerbated, if not brought on, by well-meaning men, men who are in many respects "larger than life" or who might have seemed so, at least, until greater forces came along. In this instance, then, the story of the building of the Key West Extension of the Florida East Coast Railway and its ultimate disposition fills Aristotle's bill—it is, in fact, tragedy incarnate.

Today, US Highway 1 is alternately the bane and the lifeblood of the local populace. It is the only way in and out of the Keys for most, and when an eighteen-wheeler jackknifes and comes to rest across those two narrow lanes, or a drunk plows head-on into a counterpart hauling the other way, chaos results. For hours, even a day or night, depending upon the degree of carnage, no one goes into the Keys, and no one gets out.

Some of the locals (you can call yourself a "Conch" if you were actually born in the Keys) lament the situation, but it's a lot like complaining about the weather. There has been talk of widening the road to four lanes, but that would bring its own set of problems: aside from the argument that more lanes will simply generate more and bigger accidents, there is the havoc certain to be wreaked on an already fragile environment, as well as the specter of encouraging further development where more than enough has barnacled itself already.

On the positive side, the highway permits speedy travel north for Keys residents, a release valve for what they sometimes call "island fever." Along the hundred miles of highway down the Keys, there are no malls, as most Americans know them. You can count the number of movie screens between Homestead and Key West on both hands, the number of bookstores on one. Yet a distance that would otherwise take the best part of a day by boat (and not so long ago did) can be traveled these days in three hours, tip to tip.

And from the mainland down US 1 (there were no designations, by the way, before the National Highway Aid Act of 1925 replaced "named" highways such as the Tamiami Trail or Dixie Highway with a numerically based, east-west/north-south grid known as the "Uniform System") flows everything Keys residents need to live: food, medical supplies, lumber, new video releases, beer. Even fresh water, for there is no natural source of it in the Keys, and the single big pipe that carries that elixir for every resident from Key Largo to Key West follows right alongside the roadbed.

For the tourist, moreover, there's little downside to US 1. That's how most of them reach America's southernmost point in Key West—and all of the glittering marinas, the upscale resorts, the dive spots, the bonefish flats, and the jaw-dropping marine wilderness vistas scattered along the Keys in between.

There's a commercial airport in Key West and an executive airstrip in Marathon, and sailors still make their way by boat along the Intracoastal Waterway and from every port in the

world, of course. But by far the greatest numbers of visitors to
this singular part of the world come to it by road.

And what a road it is.

The first few miles of US 1 out of Homestead, an agriculturally based city of some twenty thousand, are, save for their
utter desolation, comparatively unremarkable. Homestead, it
might be noted, is the place that took it squarely on the chops
from Hurricane Andrew in 1992. Though the storm leveled
most of the area, chasing the Cleveland Indians out of their
new spring training camp, driving a stake through the heart of
Homestead Air Force Base, and running up the largest storm
damage tab in American history, locals still shake their heads
and wonder what the devastation would have tallied had the
hurricane strayed northward thirty miles or so to hit Miami
and Miami Beach square-on.

On a pleasant day, though, the concerns are less dire. LAST
CHANCE FOR BEER AND BOOZE warns a sign that flaps above
a battered saloon and package store on the far-south fringes of
Homestead, at journey's outset.

And it is not an idle threat. From here the road bores southward into the virtual nothingness that marks the verge
between Florida's mainland and the Keys.

The narrow two-lane is bordered first by a mile or so of
feathery Australian pines—tall, trashy intruders from another
continent, which fall away gradually, offering unbroken views
across unpopulated swaths of saw grass, interrupted here and
there by dark hammocks of mahogany and scrub—hallucinatory vistas not unlike those across the baking African veldt.

There's a gravel spur spinning off the highway here and
there, leading off toward a distant rock pit, a rumored work
camp, or perhaps a secluded turnaround where lovers meet or

stolen cars are stripped and dumped. But as the sign promises, there is nothing truly civilized: no houses, no pit stops, no gas pumps, no cold beer to be found.

In these parts, the most remarkable feature is actually to be found high up in the cross-timbers of the huge electrical pylons that flank the highway, where the great ospreys—the eagles of the sea—have discovered an agreeable place to build their nests. These are massive tangles of interlocked tree limbs and driftwood, a dozen feet or more across, half again as tall, draped over the poles' crossties and braces—and protected from human molestation, by the way—floating up there like beaver dams tossed up by an apocalyptic tide. Since there are no trees within hundreds of miles anywhere near the size of the pylons, drivers might wonder how the ospreys made out before progress came along. It's the sort of idle speculation a drive through such country encourages.

Halfway to Key Largo, though, the terrain shifts again, the road tunneling now through an unbroken wall of mangroves, a gnarly tree that rarely grows to more than the height of a Greyhound bus, and within whose watery roots are sheltered the fry upon which most of the Florida fishing industry is based. Water in roadside canals laps at the highway's shoulders here, and rare gaps in the mangroves offer a tantalizing glimpse of the Atlantic on one side, the Gulf of Mexico on the other.

For the most part, though, it's fairly claustrophobic, even monotonous, travel just south of Homestead—and some of the worst accidents in Keys history have taken place here. Drivers find it easy to nod off in this leafy tunnel and veer into the opposing lane, and even the most defensively oriented don't have much maneuvering room. To make matters worse, rescue workers often find themselves stymied in trying to reach crash scenes; EMS vans have a difficult time weaving between miles of stalled traffic on the narrow track, and it's even ticklish trying to bring a helicopter down into such tight quarters.

"It's about the last place on earth I'd want to get hurt bad," says Alvin, a Miami paramedic and my new acquaintance.

We're standing in the middle of the highway at a spot near the end of the Homestead–Key Largo leg of US 1, between the back of his pickup and the nose of my car, waiting for the traffic southward to start moving again. It's not an accident that has caused this pile-up, however. It's just that the drawbridge at Jewfish Creek is up, and in a few minutes, as soon as the tall-masted sailboats and the larger, motor-powered "stinkpots," as the sailing purists like to call them, have made their way through the cut that separates the normal world from the Keys Republic, we'll be on our way.

Spending any significant time in the Keys makes it clear that it's not an unusual occurrence to make friends this way—there are tales of keg parties and impromptu dances in the middle of crash-closed bridges—but try to imagine the same thing happening during a snarl on the FDR Drive or the Hollywood Freeway, and one gains the first inkling of what lures people down this way.

Soon enough, the bridge is inching downward and Alvin is headed back toward his pickup. "Make sure you pull over to the side if you have to write something down," Alvin cautions, using the tip of his Budweiser longneck as a pointer. If he's aware of any irony in the gesture, his earnest gaze does not belie it.

There's a whine beneath the tires as you cross the steel deck of the Jewfish Creek drawbridge, and, at long last, a serious look at real water for the first time as the road bisects a mile or so of Lake Surprise, so named for the reaction of the first railroad-building crew when they hacked their way down to this point from Miami. If your windows are down, you can get a good draft of sea breeze here as well: salt, seaweed, ammonia-tinged air. It's the signal that things have changed, though it may be a little while until it's clear just how much.

At this point a motorist may also take notice of the "mile markers," or MMs, little numbered signs that have begun to pop up along the shoulder of the road. Among the first is MM 107, near where US 1 converges with Card Sound Road on Key Largo, meaning that there are 107 miles of highway between that point and the end of the line in Key West. The mileage counters form the basis of all Keys addresses from that point on.

Key Largo is by far the largest in the chain of islands, and the nearly twenty miles of highway that run along its coral spine could be laid along the outskirts of some Midwestern city: strip malls and geegaw shops abound here. The town itself was called Rock Harbor until 1948, when local business leaders saw the wisdom of cashing in on the popularity of the Bogart-Bacall movie that used the interior of a local roadhouse for a few atmospheric scenes.

A few miles farther along, a motel has installed the river-boat used in the filming of *The African Queen* out front, touting Bogie's long-standing association with the area. Both movies were essentially studio productions—one in Holly-wood, one in England—and so far as anyone knows, Bogart never set foot on Key Largo, but it's a diverting notion to think otherwise, clipping off the miles toward Tavernier, the first of the truly tiny Keys settlements along the way.

At MM 90 is Plantation Key, upon which the town of Tavernier sits. The Key is three miles long, and little more than a couple of football fields wide. The smell of the sea is stronger here, though there's still enough development and foliage to keep the water hidden. There is little sense at all of the loveliest feature of the area: Pennekamp Coral Reef lies just offshore, the nation's first underwater state park, and probably the premier dive spot in American waters.

Through Plantation and Windley Keys, the road courses through another dozen mangrove- and motel-lined miles, past an abandoned quarry converted now to something called Theater of the Sea, where dolphins, sharks, and sea lions are promised to display themselves, past a massive marina-cum-bar and lodging complex called Holiday Isle, which might seem, with its hordes converging on it, to have been the end point of this highway . . .

. . . but in short order the throngs are left behind, the road suddenly heaves upward and becomes airborne, arching high over the channel that connects the Gulf and the Atlantic here, and for the first time the traveler understands that, indeed, this is a singular part of the world.

It's an osprey's-eye view here at MM 84, out over the patchwork-colored seas. Splashes of cobalt, turquoise, amber, beige, and gray alternate, then fall away to deeper blue and steel, and off toward a pale horizon where sky and water meet at a juncture that's almost seamless on the brightest days. The variegations of color have to do with the time of day, the cloud configuration, the nature of the sand or grass on the sea bottom, and the shifting depths of the water itself surrounding the Keys, which can range from a few inches to a few feet, and then plunge several fathoms and back again in an eye-blink.

To the west, there's a view of a key called Lignumvitae, three-hundred-odd acres constituting a true island that rises sixteen feet above sea level and where giant mahoganies were once logged, today the site of the last untouched tropical forest in the state. To the east is a much smaller dot of land called Indian Key, once the outpost of a gentleman pirate, or wrecker, named Housman and the staging area for John James Audubon, who came to the Keys to shoot and then sketch tropical specimens for *The Birds of America*. Not much remains on Indian Key, but its modest aspect probably accounts for the urge that begins to creep into the back of the traveler's mind at about this moment:

. . . desert island, private island, island paradise. Buy myself one of these little dots, get a boat, and build a dock, kiss the world good-bye . . .

There are not many places where such unspoiled islands still exist, not in so dramatic a setting, anyway. It's an atavistic urge, perhaps, but one that likely pulls the traveler resolutely southward now.

There is civilization to be encountered once the highway descends again, to Matecumbe Key, but except for the occasional exquisite resort (Cheeca Lodge and Hawk's Cay among them) tucked away behind the palms, it has a decidedly temporary look about it: restaurants, motels, and houses that seem hastily assembled, with little expectation of their staying long. One of the world's most exclusive sporting resorts—the Long Key Fishing Club—once resided here, but the hurricane of 1935 swept it away. When the storm had passed, observers found little trace of the club. Even today, historians argue that a memorial erected to mark the site actually stands far from where the club was built.

Below Matecumbe, everything changes markedly. And while there is still land, bits of it, anyway, stitched together by a ribbon of highway, this is truly Water World.

South of MM 65, two and one-half miles of multihued water separate Lower Matecumbe from Conch Key to the south, a stretch once crossed by a railroad bridge, the remnants of which still stand as a fishing pier stretching, unspliced, from the south and north, and paralleling the modern highway that zips alongside. Water everywhere, right, left, ahead, and behind, so much of it that land is nearly out of sight; fishermen standing a few feet away, vying with pelicans for elbow room on the old bridge rails: the traveler might wonder whoever had the thought to build such a road as this in the first place.

It's a thought that won't go away, despite another interlude of landlocked driving through Grassy Key, a spot so isolated that big-city aquaria and water theme parks regularly send their overstressed dolphins to the marine research center here for a little R and R. After Grassy Key comes Crawl Key, and next Key Vaca, and the sudden quasi-urban sprawl of Marathon, where actual airplanes are tethered along the narrow runway at highwayside. There is a plethora of motels and restaurants in Marathon, most of them there to service the moneyed sport fishermen who have come to pursue the more than four hundred varieties of fish that are said to live in the waters surrounding the Keys.

Yet this overbuilt enclave marks another stage in civilization's steady unraveling down the archipelago. When the original railroad was being built along this path, progress was swift and steady, despite the many obstacles this unique terrain presented. Even the Long Key Viaduct was built rapidly and without incident, relatively speaking.

But just past Marathon, at MM 47, land as most people reckon it truly ends. From this jumping-off point stretches seven miles of open water. Even the railroad builders were stymied. For three years, engineers struggled and men died to span the unthinkable distance between Knight's Key and Little Duck, until finally it was done.

Following the railroad's disappearance, the highway was built atop the old railroad span, and in the 1980s, another modern bridge was built alongside. As was the case with the original Long Key Viaduct, much of the original railroad bridge was left standing, some of it serving as fishing pier, some of it simply remaining, massive stretches jutting up from the water, pilings and arches built literally out of sight of land, as obdurate and mystifying to the modern traveler as the vestiges of Stonehenge:

"What *is* that over there, anyway?"

"Old railroad bridge."

"Railroad?"

"Yep."

"Across the ocean?"

"That's what it is."

"Who would build a railroad across the ocean?"

"Now that's another story."

A lucky traveler might get that much out of a typically closemouthed Conch, maybe one he'd bumped into on the rocky beach at Little Duck, MM 40, or Missouri or Ohio Key. But there might not be anyone on those flyspecks of land, not unless someone had come to do a little roadside fishing or load up a pile of the lobster traps that are often stored along this lonely stretch of road.

Along with Bahia Honda or "Deep Bay" Key, this relative hiccough of land marks a true geological distinction between the Upper and Lower Keys, the Upper Keys being formed primarily of ancient coral, the lower an upheaval of limestone that nourishes a somewhat wider range of plant and animal life. Separating the two are the waters of the Bahia Honda

Channel, the deepest to be encountered in the Keys, and a fresh challenge to the builders of the railroad.

While the waters to be crossed at Long Key and Seven Mile were vast, they were at least shallow. Pilings could be sunk in water a few feet deep at most. And, as engineering science holds, even hurricane-driven waves could not logically exceed the depth of the water, so the height of the bridges was correspondingly modest.

At Bahia Honda Channel, MM 38, engineers encountered steely blue waters, however, and divers soon confirmed their fears. "Twenty-three feet in some places," was the report. "Thirty-five in others."

Which meant the pilings that were to hold the rails had to rise up in response, or else risk railroad passengers being swamped at sea, a concept as bizarre as it was terrible. So progress slowed to a crawl while the railroad builders could once again devise a way to do what had never before been done, all the while mindful of the advancing age of Henry Flagler and his determination to ride his own "iron" to Key West before he died.

It is the sort of knowledge that few contemporary drivers contemplate of course. Those lucky enough to have timed the trip with the onset of evening might slow down to savor the view from the soaring bridge. Others, as desperate in their own ways as Flagler to reach Key West, find the bridge's downhill slope a welcome spur to pick up speed.

On the other side of Bahia Honda, the low-lying keys of Big Pine and Ramrod, Cudjoe and Sugarloaf, are notable for disparate, even bizarre, reasons. Cudjoe, for instance, is home these days, to "Fat Albert," an unmanned interagency surveillance blimp that floats high above the island like a perpetually tethered cloud, keeping its ultra-high-tech electronic eyes and ears attuned to all naval goings-on—drug running not the least of it—in the Caribbean corridor.

At MM 31, Big Pine is home to what's left of the herd of tiny Keys deer: creatures not quite the size of a Great Dane and whose near extinction in the mid-twentieth century was one of the spurs to the passage of the Endangered Species Act. There is some debate among scientists as to whether or not the Keys deer are a native species all their own, or are the descendants of the Southern whitetail deer scaled down to Keys size by eons of hardscrabble life in a circumscribed environment, but in any case they are remarkable to see, assuming you can find one. Though the numbers of deer have increased in recent years, it's not as though you're likely to find the shy creatures grazing in herds along the road. And for creatures that are seldom seen, it's the long-sought-after residents of the Bat Tower, on Sugarloaf Key, that are hard to beat.

To reach this attraction, a thirty-five-foot wooden colossus that looks something like a mine shaft housing or, well, a bat tower, requires a short detour westward off the highway. The structure was built in 1929 by Sugarloaf's original developer, R. C. Perky, a man who'd hoped to somehow attract a colony of bats that would in turn eat the clouds of mosquitoes that favored Sugarloaf, and that were discouraging Perky's efforts to lure tourists to his island. Despite the deployment of a top-secret bat bait sold to Perky by a Texas entrepreneur, no bat has ever lived in Perky's tower so far as anyone knows, but the structure remains, a somewhat lesser testament to the grander dreams of man.

The Bat Tower, near MM 17, may be the last sight of interest to divert the resolute traveler from the jewel at the end of the road. From there, Key West lies less than twenty minutes' drive to the west and slightly south, with only Big Coppit (not big at all), Boca Chica (home to a Navy air base), and Stock Island (home to Mount Trashmore, the highest point of land-fill in the Keys) to intervene.

Key West is, after all, the point of this journey for most, as it always has been, as it always is likely to be.

There's something special about Key West. It is the closest habitation the United States has to being truly tropical, lying fewer than fifty miles north of the Tropic of Cancer, and as locals are fond of pointing out, it is far closer to Havana (one hundred miles) than to Miami (half again as distant). Its geographical position is, in fact, one of the points that Henry Flagler raised when asked to explain his otherwise unfathomable urge to build a railroad here.

But Flagler's railroad across the ocean never earned a dime of profit, and it is difficult to imagine how a businessman as bright as he was ever thought it would. Flagler managed to fabricate excuses for his endeavor, one of them being that some would come to call a World Wonder. And while tourists made use of the line, freight shippers—the bread and butter of the railroad business—never did.

Plenty of practical excuses for Key West have been dreamed up over the years—military garrisoning, cigar making, sponge diving, shrimping, turtle raising, pirate sheltering, drug and booze running, and Flagler's speculative notion of a deep-water, South American gateway port among them.

But the truth is that Key West has survived in spite of all these practical notions, principally by providing pleasure in a dizzying array of forms. Fishermen and artists, divers and drinkers, day-trippers and dropouts, Navy men and those who ogle them, presidents and paupers: Key West has lured them all. (Harry Truman once maintained a southernmost White

House on the grounds of the naval station there.) Key West remains to this day beyond practicality, the ultimate destination, a rocky island whose siren song has lured so many to its shores, even the no-nonsense partner of John D. Rockefeller himself.

3

Citizen Flagler

In early 1904, when Henry Morrison Flagler made his fateful decision to begin the building of the Overseas Railroad, he was already seventy-four, and, in the eyes of most, was nearing the end of a second successful career. Certainly the drive to make money had little to do with his decisions in those days, even if money, or the lack of it, had been the central force in the first part of his life.

Flagler had grown up poor, the son of a Presbyterian minister with a parish in northwestern New York State. Young Henry was only fourteen when the family's spartan existence prompted him to leave home in 1844 and join his half brother, Dan Harkness, in northern Ohio, for a stint as a salesman in an uncle's general store. Flagler, who arrived in Bellevue with a few pennies in his pocket, was determined to make the most of his opportunity, working long hours to save his money and often refusing invitations to join acquaintances on weekend jaunts to nearby Sandusky. His hardworking, sobersided ways would persist through much of his life, earning him the trust of employers and, later, of influential investors and partners who would change his life beyond his dreams.

Part of the ambitious young Flagler's duties came to involve the brokering of local corn to shipping agents in nearby Cleveland. Though he knew nothing of the grain business at the outset, he threw himself into its study with his characteristic devotion to the job at hand. His singleminded approach was so successful that he was able to buy into the Harkness family business within a few years, and shortly afterward made the acquaintance of one of his Cleveland counterparts in the grain-brokerage chain, one John D. Rockefeller.

Though Flagler's upbringing had been puritanical and he himself was a virtual teetotaler at the time, one of the natural adjuncts of a grain dealer's business was the maintenance of a distillery, a sideline that constituted a ready conduit for the use of surplus grain. Faced with a choice between his scruples and his overriding desire to succeed, Flagler barely wavered. It wasn't long before the distillery business was as important to him as the merchandising of corn.

The onset of the Civil War proved a boon to Flagler, who, though he was opposed to slavery, saw no reason to go to war over such matters. While Dan Harkness, now his partner in the grain brokerage, volunteered and went off to fight, Flagler stayed home to tend to business, encouraged in his decision by Rockefeller and others of their circle who felt that the war was a distant and wasteful distraction.

A distraction, perhaps, but certainly a profitable one for a grain merchant who began to realize the truth of the maxim that an army travels on its stomach. Business boomed, and Flagler was soon rich by his own standards—rich but bored. He had fifty thousand dollars in his bank account, a tidy sum in 1862, and a king's ransom in the tiny town of Bellevue, Ohio.

Casting about for something more interesting to do, Flagler hit upon the idea of . . . salt. Intrigued by the discovery of vast deposits of the mineral in nearby Michigan and an act of that state's legislature that made the business tax-exempt, Flagler sank every penny he had into the venture, along with an equal amount that he borrowed. But the great salt rush had drawn a horde of competitors, some of who actually knew a few things about the business.

When the end of the Civil War brought a collapse in prices, Flagler's operation fell apart. He found himself not only penniless, but fifty thousand dollars in debt. It was a lesson the ambitious young man would never forget: failure was simply not acceptable.

He returned to Bellevue from the offices he'd been keeping in Saginaw, Michigan, and borrowed enough money from his relatives to satisfy his creditors. Though he could have opted for safe haven there, Flagler was resolute. He might have been beaten, but he would not move backward.

Instead, with a few hundred dollars in his pocket advanced him by his father-in-law, he moved with Mary, his wife of eleven years, to Cleveland and renewed his old acquaintances in the grain-dealing world, taking a post in a firm that had been vacated by his old friend Rockefeller, who had left grain for an intriguing new substance called oil.

Because of its position on Lake Erie and its proximity to the newly discovered oil fields of western Pennsylvania, Cleveland had developed over the past dozen or so years into a shipping and refining center for the new elixir, which, at the time, was still competing with whale oil and lard for supremacy as a fuel and lubricant.

Rockefeller had invested in a refining business during the Civil War, and by the time of Flagler's arrival in Cleveland, he had decided to devote all his energies to the business of making and shipping oil. Because Flagler had rented a house on the same street as Rockefeller and kept his offices in the same building, the two often walked to and from work together, comparing notes and sharing their chief, binding passion: the desire to make large sums of money.

Rockefeller was convinced that oil was the conduit to success, and he had joined forces with a chemist by the name of Andrews, who possessed the technical expertise upon which the refining process was founded. Rockefeller himself was the consummate manager. But he was well aware of his own shortcomings as a marketer, and that was where Flagler came in.

Flagler was one of the most successful grain brokers Rockefeller had known in the halcyon days before the war. At thirty-six, Flagler was nine years older than Rockefeller, and, if not handsome, was at the very least striking, with a vigorous head of hair and a full mustache, and a personality that radiated confidence and drew others to trust him.

Rockefeller valued Flagler's undying optimism and drive as well as his relative maturity, which would come in handy for a fledgling business founded upon a new technology and seeking to attract investment from others. When one of Flagler's wife's cousins offered to invest $100,000 in Rockefeller's new venture, the deal was made. It was an agreement that would alter the course of American history.

For the next fifteen years, Flagler and Rockefeller worked side by side, walking to their offices together in the mornings, passing drafts of letters and detailed business documents between their desks during the day, walking home together at night, always planning and calculating. The result of their efforts was Standard Oil, the largest, most powerful, most profitable, and perhaps most notorious corporation ever formed.

Rockefeller would come to freely attribute the secret of the firm's success to his partner, for they were not long in the business, said Rockefeller, before Flagler realized that the negotiation of a lower freight rate was the key to the entire matter. If oil could be brought to their refineries at a rate below that offered to competitors, it would create an unassailable competitive advantage. In the highly competitive oil market, no other factor in the process could differentiate one player from the next to such a degree.

As a result, Flagler soon became a master at the negotiation of rebates from the major rail carriers who served the oil fields, carrying crude oil to Cleveland for processing. In return for lower rates, Flagler would guarantee massive shipments to the railroads. To meet these goals, Flagler would in turn have to acquire more crude and increase his refining capacity, and in order to make that happen, he and Rockefeller would need money, a lot of it.

Later, a writer was to ask John D. Rockefeller if he had had the idea to incorporate the business. Rockefeller minced no words. "No, sir, I wish I'd had the brains to think of it. It was Henry M. Flagler." With Rockefeller's grudging agreement, the Standard Oil Company went public in January of 1870, its capitalization of $1 million divided into ten thousand shares. Rockefeller took about 2,600 of the shares, and Flagler about half that. All but one thousand of the shares were taken by various insiders. Inside a dozen years, the worth of the company would grow to $82 million, a staggering rate of increase, and one fueled largely by Flagler's remorseless goal to control completely the production of refined oil in Cleveland.

A reserved and devoted family man in his personal life—he was dedicated to the care of his fragile wife, Mary, and was reportedly content to spend his evenings and weekends at home by her side—Flagler was a ferocious tactician at the office. Within a few months, he and Rockefeller had either bought out or scared off twenty of their twenty-five competitors. The choice offered to most was simple: accept what Flagler always insisted was a "fair" price for their holdings, or go broke trying to compete with a powerhouse that could do business more cheaply.

In a letter to an associate of the day, Flagler displayed his characteristic approach to business negotiation: "If you think the perspiration don't roll freely enough, pile the blankets on him. I would rather lose a great deal of money than to yield a pint to him at this time."

Flagler's irksome tactics were not limited to his fellow refiners. In 1872 he took advantage of a fall in oil prices to persuade most of the Pennsylvania oil producers to join with him in a scheme directed at the entire rail industry. In what may sound familiar to those accustomed to today's OPEC shenanigans, Flagler proposed an industry-wide agreement to limit oil production, thereby guarding against price fluctuations, and also forcing rate concessions from carriers who would have to play ball or be frozen out.

Though public outcry foiled the most egregious of these "associations," sub-rosa agreements of the sort were the order of the day. And, fat with the ever-growing profits, Standard Oil could afford to construct its own transportation systems, including the newly developed network of pipelines. The rich simply got richer.

By 1877 the company had become a behemoth that had far outgrown its Cleveland roots. Rockefeller and Flagler determined to move their operations to the burgeoning city of New York, where the company's far-flung holdings could more easily be managed and where other titans such as railroad builder Cornelius Vanderbilt, fur mogul William B. Astor, and department store maven Alexander T. Stewart had made their homes.

Despite the heady move, Flagler was not keen to join the New York City social swirl. Even in Cleveland, he had virtually no social life. His wife had been plagued by a lifetime of chronic bronchitis, and when Flagler was not at his office, he was with her.

The move to New York did little for Mary's health, and when her doctors changed their diagnosis to tuberculosis, they also suggested that a winter in Florida might improve her condition. Flagler did not hesitate. Despite the pressures of a massive business and the growing antitrust fire directed at Standard Oil, Flagler accompanied his wife and their son and daughter on a train as far south as Jacksonville, where, as history would again note, a lack of adequate transportation and a dearth of decent accommodations halted the entourage.

Mary responded well to the balmy climate, however, and the Flaglers would return to Jacksonville again, though she was hesitant to stay long once her workaholic husband had returned to the fray in New York City. In spite of the forays to Florida and the best of medical care, which Flagler's wealth provided, Mary's condition continued to deteriorate. By the winter of 1880 she had become so ill that doctors advised Flagler to cancel their planned return to Jacksonville. Mary's condition continued to worsen, and in May of 1881, she died.

Her death was a stunning blow to Flagler. The Flaglers' twenty-six-year-old daughter, Jennie Louise, was married and living with her husband, but with his eleven-year-old son, Harry, still at home, Flagler resolved to do a better job at fatherhood. Despite the irony for a preacher's son, Flagler bought a grand estate named Satan's Toe in Mamaroneck, forty-two rooms on thirty-two acres of land overlooking Long Island Sound, and persuaded his half sister, Carrie, to come live there and help tend to his young Harry.

To prepare properly for this new phase of his life, Flagler saw to the absolute renovation of Satan's Toe, including the installation of fixtures he picked out himself, the building of a two-hundred-foot breakwater, and the construction of a sandy bathing beach along the shore. Satan's Toe had been transformed into a resort destination that was the talk of New York society, and the fifty-two-year-old Flagler, for the first time in his life, was taking pleasure in something that did not have to do with work.

Concurrent with these new interests had come a withering condemnation of Flagler's business activities from the newspapers, the public, and governmental agencies alike. While a country torn asunder by the Civil War had been all too happy to see prosperity return to the nation during the 1870s, it was not long before a feeling of laissez-faire, if not outright gratitude, directed toward successful business interests was replaced by scrutiny of those same organizations.

Competitors who had been steamrolled by Flagler and Rockefeller had for years complained bitterly about the high-handed tactics of the all-powerful Standard Oil, and several investment groups had been formed to build pipelines of their own that would compete with Flagler's virtual lock on the rail transport of oil. When Flagler began to call in political favors to block the new competitive threats, public resentment reached a crescendo.

In December of 1882, Flagler was called to testify before a Senate antitrust committee in New York, where he was badgered unmercifully by the Senate's attorney, who insisted that Flagler stop hedging and answer his questions directly. As tensions mounted, an intransigent Flagler shouted at the man, "It suits me to go elsewhere for advice, particularly as I am not paying you for it."

"And I am not paying you to rob the community, I am trying to expose your robbery," responded the attorney.

The confrontation ended in deadlock, though Flagler had begun to see the writing on the wall. Entire national political parties—Greenback, Union Labor, Prohibition—were being based largely upon antitrust platforms, and a number of the larger industrial states were in the process of enacting monopoly-busting legislation.

It was not so much that the value of Flagler's holdings was threatened: by 1888 the value of Standard Oil shares had risen to more than $150 million, and even the eventual dissolution of the Standard Oil Trust was similar in financial impact to the reorganization, nearly a century later, of Ma Bell into all those little Baby Bells. For anyone who held substantial stock in a parent company of such clout, money was hardly an issue.

Flagler had amassed a fortune, it is true, but at the same time his monumental business achievements had brought him the apparent enmity of an entire nation. In addition, he had lost his wife, the virtual supporting pillar of his private life.

It should have come as no surprise, then, that a man in Flagler's position—wealthy beyond imagination, his public life a source of never-ending condemnation, his personal life virtually obliterated—should be poised for a sea change.

But Flagler had begun to see the possibilities of satisfaction to be derived from sources outside the arena of business. His renovations at Satan's Toe, along with the favor it found in the eyes of friends and business associates, had provided him with unexpected pleasure, and another uncharacteristic force had entered the purview of his life as well.

One of the nurses who had attended Mary Flagler during her final years was a young woman of thirty-five named Ida Alice Shourds, an attractive woman with flaming red hair, bright blue eyes, and a volatile temper. Flagler, who had lived thirty years with an attractive though restrained and often bedridden mate, found himself smitten. Though Ida Alice had no formal education and did not share Flagler's own enthusiasm for reading and modest cultural interests, neither that nor the general disapproval of his friends and family seemed to bother the fifty-three-year-old Flagler greatly. His courtship was as resolute as his acquisition of wealth had been; in June of 1883, Flagler and Ida Alice were married.

Even Ida Alice's fabled shopping sprees, which netted her one of the most elaborate wardrobes in New York City, didn't faze him, for Flagler was a wealthy man, his net worth at $20 million and climbing with every barrel of crude oil that the vast Standard Oil combine pumped out of the ground. He had made it beyond his wildest expectations, this poor, puritanical boy from the sticks, and it seemed that he was ready to enjoy the fruits of his labor at last.

Paradise Found

4

Though Flagler, then fifty-three, and Ida Alice, thirty-five, had married in June, business considerations delayed their honeymoon until December. To escape the frigid New York weather, Flagler proposed a return to Jacksonville, where he and Mary had spent some of their more pleasurable days.

Flagler and his new bride traveled by rail to Jacksonville, and, after a few days rest, embarked on a sail down the St. Johns River to the historic town of Saint Augustine, founded by Spanish explorers in 1565 and the oldest settlement in the United States. To the Flaglers, harried by the press of city life, and mindful of a subzero cold wave that gripped the North, balmy St. Augustine, with its two thousand inhabitants, waving palms, and blooming orange groves, seemed like paradise itself. The honeymoon would last until March, and less than a year later, Flagler and Ida would return.

This time Flagler combined business with pleasure. He'd heard that a new hotel was being built in St. Augustine, and had been developing ideas of his own. He met with a Boston architect who had built a winter home of his own in St. Augustine, using a new poured-concrete process that allowed for considerable fluidity in styling, even where larger structures were concerned.

It was the key that Flagler had been looking for. In short order he had bought up a large section of unproductive orange groves, had hired himself a New York architect, and embarked upon the building of a lavish Mediterranean-themed hotel—the Ponce de Leon—in St. Augustine.

News of Flagler's project swept through New York City financial circles like wildfire. For a man whose stature was the equal of John D. Rockefeller's to embark upon a project of such magnitude at that time would be a bit like Walt Disney announcing plans to build a second version of Disneyland in central Florida. The ensuing publicity set off the first of the great Florida real-estate speculation frenzies, and Flagler was deluged with offers from every quarter, most of which he brushed aside.

For while this was clearly business, it was business of a different sort. No degree of success in hotel management could ever provide an income rivaling what had come from oil.

During an 1887 interview granted to the *Jacksonville News Herald*, Flagler was asked to explain why on earth a man with a major interest in the most powerful company on earth would want to get into the hotel business. Flagler responded by telling a story he'd grown fond of—that of the elderly church deacon asked to explain a sudden, unaccountable bout of drunkenness. The deacon explained to his pastor that he had spent all his days hitherto in the Lord's service, Flagler said, and now he was finally taking one for himself.

Similarly, Flagler told the reporter, "For the last fourteen or fifteen years I have devoted my time exclusively to business, and now I am pleasing myself."

Later in the same interview Flagler elaborated: "[T]he Ponce de Leon is an altogether different affair. I want something to last all time to come. . . . I would hate to think that I am investing money that will not bring a return in the future. I will, however, have a hotel that suits me in every respect, and one that I can thoroughly enjoy, cost what it may."

Those were fateful words, spoken at the beginning of a new career for Henry Flagler. It was a credo that would inform all the subsequent work of "the man who built Florida," a two-decades process that would culminate in the "lunatic notion" to build a railroad across the sea.

It is doubtful that Flagler was looking that far ahead at the time, for the building of the Ponce de Leon was no easy task: at the time, malaria was an uncontrollable threat to workmen in St. Augustine's hot and humid climate, and the amount of native rock that needed to be quarried from nearby public lands required a government waiver. It took nearly a year and a half to build the 540-room hotel, a process that Flagler himself oversaw, down to the opening of crated furniture alongside his crews.

One of the favorite stories retold by Flagler biographers concerns the builder's predilection for dropping by the building site unannounced, to see how things were going. One day, however, Flagler, who was smoking a cigar, found his way blocked by a zealous guard. The guard pointed out one of the many "No Smoking" signs posted about and furthermore informed Flagler that there was to be no trespassing on the construction site. When Flagler protested that he was the owner, the guard was unfazed. There had been a good many Flaglers showing up all week, trying to get a look at what was going on, the guard announced, and he'd thrown every one of them out. Flagler was still trying to talk his way in, when one of the general contractors happened by and began to chastise the guard for not recognizing whom he was talking to. Flagler interceded, however, pleased by his workman's steadfastness and efficiency.

All of Flagler's careful oversight was to pay off, moreover. When the Ponce de Leon opened, the national press proclaimed it superior to hotels such as Chicago's Palmer House and San Francisco's Palace, and socialites flocked southward to experience this Babylon, where even the meanest room featured electric lights and had cost one thousand dollars to decorate.

Certainly part of Flagler's success resulted from his insistence that no detail be overlooked in creating an atmosphere of splendor for his guests. But in retrospect, his timing was perfect as well.

It was, after all, the middle of what historians have termed the "Gilded Age" of American history, a period that stretched from the end of the Civil War until the Great Crash of 1929, marked by unbounded industrial growth and prosperity, and an optimism that was barely dimmed by World War I. The term was coined after the title of a Mark Twain novel published in 1873, a characteristically dark satire attacking the pitfalls and excesses of land and business speculation rampant during the postwar years.

Twain's reservations notwithstanding, public confidence and personal wealth were growing at an unprecedented rate. Centuries-old assumptions concerning the very nature of creation and man's place in the universe had come into question as well: the work of Charles Darwin and advancements in the modern science of psychology reflected a growing sense of self-determinism that had begun to work its way into the very weave of Western civilization.

The public read of the accomplishments and vast accumulations of such men as Vanderbilt and Astor, Rockefeller and Flagler, and saw no reason why they could not do the same. The works of Horatio Alger, in which poor but honest boys succeeded by dint of hard work and other Boy Scout–like qualities, were runaway bestsellers of the time; in all, Alger would write more than one hundred books, of which more than 20 million were sold.

Most important, an extraordinary number of Americans were finding their efforts and their speculations rewarded. With fortunes growing and personal income at an all-time high, the demand for ways in which to dispose of wealth grew accordingly. If Mrs. Benjamin Harrison had traveled to St. Augustine along with Vice President and Mrs. Levi Morton to hobnob delightedly with the Wanamakers of Philadelphia at the Ponce de Leon, then there was no reason why others on the swelling rolls of the Social Register should not follow.

A writer for *Harper's* visited the Ponce de Leon and, in an article titled "Our Own Riviera," wrote vividly of what he saw: "[A] woman and her lady friend and maid were paying $39 a day for rooms and meals; where an Astor and his bride had paid the same sum per day during a week of their honeymoon; where one lady took a room solely for her trunks at $10 a day. . . . There was one little party that occupied three bedrooms, a bathroom and a parlor, taking up a whole corner of the hotel on the ground floor, whose bill . . . might easily have been $75 a day. . . ." All this at a time when a skilled carpenter or tradesman might earn two dollars a day for his labors.

The response to the Ponce de Leon was so enthusiastic, however, that Flagler was soon at work on a companion hotel nearby, the Alcazar, where he intended that guests of more modest means could experience something of the sybarite's lifestyle. Flagler and his new wife took a suite at the Ponce de Leon, meanwhile, and announced that Florida was now their permanent winter home.

For a time Flagler basked in pride, having created an entity that went well beyond the practical matters of fueling kerosene lanterns and lubricating machines. He had fashioned a pleasure palace in the midst of a lone and distant place, and the result, if not exactly profitable at the outset, had taken the social world of which he was a part quite by storm.

Flagler's happiness was soon to be tempered, however. His daughter, Jennie Louise, by then thirty-three and married to the son of a Chicago industrialist, fell ill of complications during an unsuccessful childbirth. On her way by ship to St. Augustine, where it was thought she might rest and regain her health, she died.

Flagler was crushed, but he had learned something from the building of the Ponce de Leon. In a small way, he had become a creator instead of an accumulator, and had found a more substantive sort of satisfaction in such accomplishments.

As a result, he undertook to build a church in memory of his daughter and her stillborn child, a visible and positive symbol of his affection. The Memorial Presbyterian Church, constructed in the neo-Renaissance style popular of that day, remains to this day one of St. Augustine's major architectural landmarks.

The loss of his daughter was not Flagler's only concern by this time, however. Ida Alice, who had always felt snubbed by Flagler's family and his social circle, had begun to act in an increasingly irrational way. She had always been prone to fits of temper, but now the slightest irritation could send her into a frenzy. During a yachting party that she was hosting off the coast of New England, a storm came up. Though the members of the party, as well as the captain and crew, begged Ida Alice to return to port, she was adamant that her party would go on. The ship was driven far out to sea by the storm, where it wallowed for hours before making it back to port.

Such incidents troubled Flagler greatly, but he was nonetheless going to great lengths to try to make Ida Alice happy, going so far as to commission a new 160-foot yacht, the *Alicia,* and building them a permanent winter home near the Ponce de Leon, which one writer termed "worthy of Versailles."

Still, nothing seemed to appease Ida Alice, who, in an apparent attempt to garner notice, if not approval, took to scheduling an unbroken series of balls and other functions at their winter home, often appearing in increasingly risqué dress. Even a man as wealthy as Flagler had begun to feel the effects of his wife's prodigious spending. For the first time he was forced to sell off some of his Standard Oil stock, and in a vain attempt to temper his wife's enthusiasm, he began to withdraw from the endless bouts of partying.

If she missed Flagler's company, Ida Alice did not say so publicly, for she had found new company in the spirits and disembodied presences with whom she had come in contact through a Ouija board someone had given her as a gift.

It was not long until she announced that the board had informed her that the tsar of Russia was in love with her, and that they were to be wed upon Flagler's death. As if that were not enough, she began to complain that Flagler was mistreating her, and the resultant gossip among their friends pained him greatly.

In desperation, Flagler asked George Shelton, a family

friend and physician, to observe Ida Alice and to advise him on any course of treatment that might benefit her. Dr. Shelton, alarmed at what he saw, called in a pair of colleagues who specialized in mental disorders, but their arrival at the Flaglers' Manhattan apartment sent Ida Alice into such hysterics that the doctors recommended she be removed from their home and taken immediately to an asylum.

Flagler grudgingly gave his consent, but traveled regularly to see his wife. His visits to the asylum seemed only to further upset Ida Alice, however, who by this time seemed utterly out of touch with reason.

Shelton urged a distraught Flagler to remove himself to Florida for the ensuing winter, where he might better regain his spirits without the specter of his disturbed wife to torment him.

Flagler went, but it was an unhappy winter for him. When he returned north in June, he found his wife greatly improved, and prevailed upon her doctors to let him take her back to the Mamaroneck home he so loved. Though the doctors were still concerned, not the least because Ida Alice had vehemently threatened Flagler's life during her breakdown, they relented.

Flagler, by now a man of sixty-six, took his wife home, resolute that this time things would be different. Though it was not long before Ida Alice was again begging for her Ouija board, and friends were advising him to have her committed once more, Flagler stood firm.

"I shall not let her leave home until it becomes absolutely necessary," he vowed, though that would not take long.

When Ida Alice attacked her doctor with a pair of kitchen shears, the matter was decided. In March of 1897, Ida Alice was removed to a private asylum in Pleasantville, New York. Flagler would never see her again.

5

Empire Building

Flagler did not want for distractions from his devastated personal life, however. Hardly had he embarked upon a career in hotel-building than he realized that transporting customers to these emporia of delight was as important a link in the process as moving crude oil to his refineries had been so many years before.

In this case, for Flagler, there was something of a carryover from his former life. All that experience in railroading was about to be put to use in an entirely different context, as he tried to make sense of one of the most chaotic rail systems in the United States.

There had been almost no railroad construction in Florida since the end of the Civil War. The aftermath of the conflict had sent most of the operators into bankruptcy and the ensuing litigation had tied up much of the state-owned right-of-way in court battles. The lines that did exist had been built without regulation and with no regard for consistency of track. Where one line ended and another began, the gauge and type of track might vary wildly. To continue on, engines and cars would have to have their wheels refitted and their axles resized. The alternative was to unload passengers and cargo from one train and reload them on another.

It was a situation that a man who had worked with peerless organizer John D. Rockefeller could scarcely comprehend. But with construction under way on the Ponce de Leon, Flagler realized he was in dire need of better transport service over the forty-mile route from Jacksonville, which was then the southern terminus of decent rail service in the state of Florida. As matters stood, to get from Jacksonville to St. Augustine required a leisurely cruise down the broad St. Johns River, then the boarding of a narrow-gauge railroad for a few remaining miles' passage eastward to the ultimate destination.

When talks with existing line owners proved fruitless, Flagler did what anyone with his resources might: he ponied up half a million dollars and bought the railroad. As the new owner of the Jacksonville, St. Augustine & Halifax River Line, his first decision was to build a bridge across the St. Johns.

The moment that the company's engineers heard of Flagler's plans, they came forward quickly, announcing that no one had ever sunk railroad support piers in ninety feet of water, the depth they would have to cross. Flagler pondered this information for a moment, then turned back to the engineers. "Cannot you build that pier in ninety feet of water, then?"

After a brief huddle, the engineers had decided. "We can," they told Flagler.

"Then build it," Flagler replied.

The result constituted railroad history, but it was only the beginning of Flagler's involvement with railroading in Florida. Shortly afterward, he bought another Jacksonville short line and extended it directly eastward to Jacksonville Beach and its environs, where he constructed a series of coal and lumber docks that made Jacksonville a major port.

With that behind him, Flagler turned his interests southward again, extending his line to Ormond Beach, where he bought a modest inn and renovated it, renaming it the Ormond Beach Hotel, adding a golf course and other amenities so pleasing that Rockefeller built his winter home across the street.

By this time Flagler was convinced he was onto something in the providing of uninterrupted train service for tourists visiting Florida. He extended the line to Daytona Beach, laying the foundation for that town, whose twenty miles of hard-packed, snowy sand beaches would make it one of the leading resort destinations in the nation.

Residents of the lands farther south needed no convincing of the value of Flagler's efforts. He was offered free land for his right-of-way, and less than a year after the railroad had reached Daytona, it had leapfrogged another eighty miles south across the palmetto-dotted scrublands to Rockledge, almost halfway down the state from Jacksonville, across the Indian River from Cape Canaveral.

Meanwhile, a competitor of sorts had cropped up. On the west coast of Florida, perhaps inspired by Flagler's notoriety, a man named Henry Plant had been buying up a series of existing narrow-gauge railroads with the stated intention of extending a line all the way from Tampa to Miami. Plant had also built a deep-water pier that transformed Tampa into an important port on the Gulf of Mexico, and by 1891 he had completed his own extravagant hotel, the Tampa Bay, which, at $3 million, considerably exceeded the cost of the Ponce de Leon, a paltry $2.5 million.

Goaded by the outspoken Plant's vow to "outdo" him, Flagler considered what he might play as a trump card. In a letter written to the *Miami Herald* many years later, Jefferson Browne, a Key West resident and onetime president of the Florida Senate, recalls being taken aside by Henry Flagler during the grand opening of the Tampa Bay Hotel. During that conversation, Browne said, Flagler first proposed to him the notion of extending his own railroad another four hundred miles to the south, all the way to Key West.

Flagler told Browne in that conversation that the logical end of all railroad building in Florida was to reach a deep-water terminus in proximity to Central and South America. In fact, some have argued, had Flagler been successful in getting the U.S. government to help him pay for the costs of dredging such a harbor in Miami's Biscayne Bay, there might never have been a Key West Extension.

In any case, the Panama Canal was sure to be built one day, Flagler told Browne, while Plant's imported symphony played and opera stars sang, and the nearest deep-water port in the United States was sure to have an enormous advantage. It soon became clear, Browne recalled, that Flagler was seeking Browne's assistance in seeing that the existing government franchises for building such a line be set aside.

In 1883, General John B. Gordon of Georgia had obtained the first franchise from the state for the building of a railroad to Key West, but Gordon had little capital of his own, and had secured the rights in hopes of attracting deep-pocketed investors to the project. After building a few miles of track on the mainland, Gordon had gone broke, but other speculators had acquired the franchise in turn.

As Flagler argued, this series of petitioners were only

schemers involved in the grossest speculation (not unlike contemporary consortiums tying up Internet domain names or seeking rights to the first Burger King franchise on Mars). Browne listened intently to Flagler, and, apparently unimpressed by the thousands of guests who strolled and rode rickshaws about the grounds of Plant's hotel—dukes, duchesses, and theater stars included—finally gave his agreement to the railroad man from the other side of the state, sensing that Henry Flagler was the one man in all creation who might be able to pull off such an impossible feat.

Warming to the advantages such a railroad would open to his constituency, in 1894 Browne wrote an essay titled "Across the Gulf by Rail to Key West," which was to be published in the *National Geographic* in June of 1896.

"Key West will within a short time be connected with the mainland by a railroad," Browne asserted, adding, "It is not too much to say that upon the completion of the Nicaragua [*sic*] Canal, Key West will become the most important city in the South."

Browne seemed to overlook the fact that the canal project, which had been mired in political maneuverings for more than twenty years, had also been assailed by critics who thought it as much of a crackpot notion as a proposal to build a railroad across the ocean. Of the latter, he was willing to grant that its having no precedent could possibly make it, "like all other great enterprises, a subject for a time of incredulity and distrust." Still, Browne asserted, "it presents no difficulties that are insurmountable."

In the piece, Browne laid out a route from Key West northward over the island chain, which he said would be protected by the neighboring Florida reef, safe from high seas "even in the severest hurricanes." If the several lighthouses that had been built along the reef had not been blown over, he reasoned, why worry about track, trestles, and bridges?

As to who was capable of building this mighty road, Browne ended his piece with another bold declaration: "The building of a railroad to Key West would be a fitting consummation of Mr. Flagler's career, and his name would be handed down to posterity linked to one of the grandest achievements of modern times."

Whether or not anyone was looking over his shoulder as he composed his article, Browne had by 1895 carried out his promises to Flagler, using his position as a state senator to see to it that all legislative impediments to Flagler's plans had been disposed of. Shortly thereafter, Flagler recombined all of his rail holdings in the state into the Florida East Coast Railway, and gave official notification to the state that it was his intention to extend his system all the way to Key West.

The proposal was so grandiose on the face of it that most lumped Flagler's Key West intentions into the same category as those of earlier speculators. He was, however, granted a charter by the legislature to extend the FEC line to Miami, a move that went beyond the mere permission to spend a great deal of money laying iron track.

An 1889 act of the Florida legislature set aside some 10 million acres of land to be deeded to entrepreneurs willing to build new railway lines and thereby bolster the state's economic infrastructure. As a result, Flagler was able to lay claim to eight thousand acres for every mile of track he built. In the end, he would control more than two million acres of land for which he had essentially paid nothing.

While sales and leasing of these lands abutting his right-of-way were a windfall, Flagler was always on the lookout for properties that might be developed as resorts, thereby creating an incentive for passengers to ride each new leg of his line. He took to riding his own railroad incognito, the better to scout out likely targets for acquisition without arousing the attention of local speculators certain to jack their prices sky-high should it be known that the great Henry Flagler might be interested.

In 1892 he had visited the seaside hamlet of Palm Beach in such a manner, and had returned to St. Augustine in a lather. "I have found a veritable Paradise," he told his managers, instructing them to acquire the necessary land for the "largest hotel in the world" and to begin planning the extension of the line to Palm Beach.

One would have to give Flagler credit for his vision. At the time the entirety of Palm Beach, situated on a narrow, palm-laden barrier island between Lake Worth and the Atlantic, consisted of less than a dozen houses. While the place was undeniably lovely, it was virgin territory, with no housing facilities for the hundreds of workmen who would be required to bring these dreams to fruition. And with no railroad line in existence, the building materials for the colossal hotel project would have to be brought down the Florida coast by a hastily assembled flotilla of cargo ships and riverboat steamers.

Although such a staging process taxed Flagler's infrastructure, it was valuable preparation for what would come later. Workmen were housed in hastily assembled communities of tents and shacks, resembling nothing so much as a vast gold-rush camp. Because Flagler was loath to mar the landscape surrounding his dream hotel, most of the camps were laid out on the west side of Lake Worth, requiring men to row to work in the mornings, then row back to camp at night. It might have been an inconvenience for those doing the rowing, but it gave Flagler another moneymaking idea.

He decided to lay out a new town on the west shores of Lake Worth, where he would erect the terminal for his railroad. It was a fortuitous decision for Flagler, but one that was to have implications that persist to this day, with the "haves" living in the palatial estates of Palm Beach itself, and the "have-nots" in what was originally conceived of as the service town of West Palm Beach. The literal distance is measured in a few hundred yards of water, but in social terms, the two municipalities are light-years apart.

Meanwhile, work on both the railroad and the hotel continued with all dispatch, workers on each spurred by a race devised by Flagler. The hotel builders won out, finishing the largest wooden structure in the world in the early spring of 1894, barely nine months after it was begun. When the 540-room structure was opened, guests might pay in the neighborhood of forty dollars a day for double accommodations, one hundred dollars for a suite.

Once again the hotel was a hit with the public, especially the moneyed public. Demand was so great that Flagler immediately commenced work on a second hotel, this set on the ocean side of the island and christened as the Breakers. Though the original Breakers structure was to be destroyed by fire a few years later, Flagler had the hotel rebuilt, and its successor remains today as popular a destination for the privileged as it was back then. Prominent among the first group of passengers to arrive, via a rail spur that Flagler had extended from the main station in West Palm Beach, were several members of the Vanderbilt family, along with a number of others on the register of the "Four Hundred," the most exclusive set of its time.

By this time, then, it seemed that everything Flagler touched would turn to gold. He was besieged with business proposals and pleas of every stripe. Hardly had he completed his line to Palm Beach than those who were aware of his charter rights were begging him to extend the rails southward to Miami, even though there was no Miami at the time.

6

The City That
Flagler Built

In the 1890s, all that existed where the modern metropolis of Miami sprawls today was a muddy settlement of fewer than five hundred souls. The place was called Fort Dallas at that time, after a long-abandoned military outpost that had been established in the 1830s where the Miami River empties into Biscayne Bay. The few hardy settlers who lived there near the turn of the century had been enticed by land speculation syndicates at work everywhere in the wild southern half of Florida, groups of businessmen who stood to make their fortunes by seeing such frontier lands settled.

Those who moved to Fort Dallas to seek their fortunes were interested in encouraging others to join them, of course. Among the most active of those pioneers was a Cleveland, Ohio, woman named Julia Tuttle, who had fallen in love with the wild but exotic setting during a visit to her father's homestead.

When her industrialist husband died and her father

bequeathed her his holdings in the area, Tuttle, then forty-one, performed an uncommon act of bravery: she pulled up stakes in Ohio and moved to Fort Dallas, intending from the outset to carve a city from the wilderness. She purchased a homestead allotment of her own from the Biscayne Bay Company—640 acres, including the site of the original fort—and went to work remodeling one of the original settlement structures into a home for herself and her two children.

Mindful of what it would take to turn the sleepy settlement into a city, she approached Flagler's rival Henry Plant about the possibility of extending his railroad from Tampa south-eastward across the Everglades to Fort Dallas. Plant dispatched his chief of railroad operations, James E. Ingraham, to investigate the 250-mile route. In what became a virtual survival march, it took Ingraham and his men nearly a month to make their way across the soggy wilderness of south-central Florida, and while they were given a hero's welcome by Tuttle and her friends, Plant, after hearing Ingraham's report, dismissed the building of a railroad across such territory, forever.

Tuttle, undaunted, turned to the other great railroad builder in Florida, offering Henry Flagler half of her land if he would only bring his railroad southward to Miami along the east-coast route. When Tuttle began her campaign, Flagler was not interested.

Though he had received a charter from the Florida legislature granting him rights to an extension of his lines to Fort Dallas, that claim presumably constituted an insurance policy for some distant future. Flagler saw no immediate reason to press his road beyond Palm Beach, not when the "city" making its blandishments was little more than a squatters' outpost.

With the grandest hotel in the world in operation and his tracks humming all the way from Jacksonville to West Palm

Beach, Flagler, now in his early sixties, felt that he had indeed reached a logical resting place. And then fortune intervened.

In the winter of 1894, one of the worst freezes in Florida history swept southward across the state, wiping out crops and citrus groves all the way to Palm Beach. The suffering he saw among farmers, growers, and laborers stunned Flagler. He sent James Ingraham, whom he had hired away from Plant, out on a private relief mission with $100,000 in cash, instructing him to disburse it all "and more, than have one man, woman, or child starve."

Flagler was also mindful of news sent to him by the indefatigable Julia Tuttle that Fort Dallas had not been touched by the freeze. Flowers still grew in profusion, and blossoms studded the citrus groves. Though legend has it that it was Tuttle who sent Flagler a bouquet of orange blossoms as dramatic proof of her settlement's favorable climate, it is more likely Ingraham who came up with that notion, following a meeting with Tuttle and others eager to entice Flagler southward.

In any case, Ingraham returned from Fort Dallas greatly impressed with the potential he saw in the land surrounding what is now Miami. And many historians insist that Flagler did have a spray of lemon, lime, and orange blossoms before him as he pondered his decision, though they say it was Ingraham and not Julia Tuttle whose idea it was.

Orange blossoms or not, the decision did not take long. Inside of three days, Flagler had made plans for what was then an arduous trip: by rail to West Palm Beach, then by launch to Fort Lauderdale, then more than thirty miles by horse and carriage to Miami.

On the journey, Flagler brought along his hotel designers, as well as his chief of railroad operations, Joseph Parrott. By the time Flagler and his men stepped down from the carriage into the balmy moonlight and gazed out over the placid waters of Biscayne Bay, it is likely that his mind was already made up. In short order, he struck the deal that Julia Tuttle had been urging upon him for years.

Flagler would receive half of Tuttle's homestead allotment and one hundred acres more, as well as another one hundred acres from the Brickell family, on the south side of the Miami River. In return, he would bring his railroad to Miami, construct another in the series of grand hotels, and develop a modern city on the site as well. Flagler confirmed the terms of the agreement in a letter dated April 22, 1895, and promised to have his railroad crossing the Miami River by the following February.

He very nearly kept his word. Construction of the sixty-six-mile extension south from West Palm Beach was aided by a sizable contingent of convict labor leased to the Florida East Coast Railway at the rate of $2.50 a month. The company had to feed and house the men, but it was still an attractive deal when private labor might approach two dollars or more per day. The process took about ten months from start to finish, and the first passenger train pulled up to a makeshift station platform in Fort Dallas in late April of 1896.

While Flagler's Royal Palm Hotel was still under construction and the rest of the town-to-be was little more than dreams sketched out on paper, the arrival of the railroad changed everything. Within three months the city had been incorporated as Miami (Flagler had to gently urge the new town council to choose the original Native American name for the settlement over his own name), and the population had soared from three hundred to fifteen hundred.

During the ensuing summer, a business section developed: a string of one- and two-story structures housing a bank, a general store, a Chinese laundry, and more, all of it resembling a mining outpost. On Christmas night of 1896, a fire broke out that leveled most of the new development, but the Royal Palm was spared.

Three weeks later the hotel on Biscayne Bay opened, five stories tall and almost seven hundred feet long, capped by an awe-inspiring lookout platform and surrounded by a golf course and palm-studded grounds. There were electric elevators, electric lights, and 350 rooms, most of them with private baths, a feature that was not common then, even at the most elegant hotels.

The Royal Palm was not only an impressive feature of the new city, but virtually the very reason for Miami's being. Flagler was not dismayed at this, of course, for over the past dozen years he had seen development thrive in the wake of his method: build a railroad to a place, erect a destination-worthy resort hotel there, and other development was sure to follow.

For a man whose former accomplishments were measured by the proliferation of belching refineries and a network of steel pipeline, watching this new process at work was bound to be heady stuff. Instead of being hounded by trust-busting government agents and muckraking reporters, Flagler found his footsteps being followed by thousands of Americans, discovering not only previously unimagined pleasures in this far-flung land, but in many cases finding new lives and careers there as well. He had little doubt that things would proceed in Miami as they had in St. Augustine and Daytona Beach and Palm Beach, but even Flagler could not have predicted the events that would cause Miami's growth to explode in exponential terms.

On February 15, 1898, scarcely a month after the Royal Palm opened its doors for a second season, the USS *Maine*, stationed in Havana harbor ostensibly to protect American interests against the incursion of Spanish colonialists, was blown up and sunk. While historians still debate whether or not the catastrophe was a put-up job, "Remember the Maine" became a rallying cry, and by April, America went to war with Spain.

Given South Florida's proximity to Cuba, the government was under pressure to put a garrison in place there to defend U.S. shores. Not a man to let opportunity pass him by, Flagler lobbied Washington to deploy a sizable number of troops, adding that Miami was "the most pleasant place south of Bar Harbor to spend the summer."

Locals, well accustomed to the annual summer mosquito plague that made life near the Everglades almost unbearable, howled at that one. But Flagler persisted. He sent a crew of men to clear some palmetto scrublands west of the settlement for a camp, and in short order the first of a contingent of seven thousand troops was on its way to Miami.

What they discovered was a town consisting of a hotel shuttered for the season and a clutch of raw buildings that might have dropped down from the sky.

In *Miami: The Magic City,* the historian Arva Parks Moore shares a letter written by one of the young soldiers: "There was a most magnificent and gorgeously appointed hotel right in the midst of a perfect paradise of tropical trees and bushes," he said. "But one had to walk scarce a quarter of a mile until one came to such a waste wilderness as can be conceived of only in rare nightmares."

The war was over by August, and the troops were soon gone, but still, Miami had become a *place* in the eyes of the world. And the outcome of the war was to reawaken in Flagler the stirrings for the last of his great projects.

7

The Stage Is Set

As early as 1891, during that fateful conversation on the grounds of his rival's hotel in Tampa, Flagler had insisted to Jefferson Browne, then president of the Florida state senate, that there was merit in building a rail line to Key West, especially if the Panama Canal were ever to become a reality. Flagler's ensuing efforts to wrest the charter away from the groups who had already laid claim to the notion were more likely the actions of a good businessman who likes to keep his options open than a signal of immediate intent to build, but the culmination of the Spanish-American War reawakened interest in the matter.

In the treaty that concluded the war, Spain agreed to give up its authority over Cuba, virtually assuring that United States interests would prevail there, a point not lost on future historians who would speculate that the scuttling of the *Maine* was an inside job, the result of a conspiracy between the U.S. government and nefarious business interests.

Whether such claims are true or not, it is certain that then-president William McKinley, elected over populist candidate William Jennings Bryan, was no foe of business. McKinley had in fact accepted an invitation from Flagler the previous year to visit the railroad baron in what he called "my domain," by which Flagler meant all of east-coast Florida, from Jacksonville to Miami.

Flagler, whose stated motivations never wavered from glowing predictions of profit, had nonetheless achieved the reputation of a fearless, frontier-busting visionary. As a report written by H. S. Duval for the Florida Internal Improvement Fund Trustees predicted, "It does not appear where the final terminus of this road is to be, but no doubt when the great capitalist learns that the Florida Keys are islands enclosed in a harbor made by natural submerged breakwater . . . and are therefore not really exposed to the violence of the outer sea . . . he may rise in a culminating spirit of enterprise and moor Key West to the mainland."

With his railroad extended very nearly to the southern tip of Florida and Miami already growing into a city, it is inarguable that Flagler had reached another plateau of accomplishment in 1898. He was sixty-eight years old, and while his investments in Florida had not prospered to the degree that those in the oil business had, he was still one of the most wealthy and influential men in the United States. Despite his familiar jest—"I would have been a rich man if it hadn't been for Florida"—he could have chosen that moment to retire and live out his years in luxury, basking all the while in the gratitude of an entire state citizenry, most of whom saw him as their champion.

But Flagler was not about to quit, not when he had come so close to the accomplishment of a goal that only a decade before had been dismissed as utter fancy. Flagler might have been in the twilight of his years, but he had spent less than a score of those years in the business of building railroads and creating cities.

Even in his late sixties, he was still a vigorous and powerful man, the white knight of an entire state of the union. He had spent considerable time and effort on the development of a deep-water port at Miami, but had encountered government resistance and had given up on that project.

He had great wealth and technical expertise at his disposal, and laid before him was an engineering task that had galvanized the minds of every professional in his field, even as its magnitude sobered the more practical-minded. As has been mentioned, it was a time in history when men were tempted no longer to regard themselves as at the mercy of the fates, but as masters of their environment. Even if such thinking would eventually prove as illusory as the philosophies that had preceded it, Flagler had no way of foreseeing it at the time.

To think of young rocket scientists at the middle of the twentieth century, staring up at the moon, equally inspired and awed at the prospect of someday reaching that destination, is not unlike a similar conjuring: a railroad engineer, fifty years before, stares out over the Straits of Florida toward Key West, filled with the same sense of wild surmise.

As Jefferson Browne wrote: "The hopes of the people of Key West are centered in Henry M. Flagler, whose financial genius and public spirit have opened up three hundred miles of the beautiful east coast of the state."

Besides, Flagler now had his practical excuse. With Spain out of Cuba, there would be an increased opportunity for U.S. concerns to forge more extensive business arrangements in that new nation.

Flagler himself had visited the island on several occasions, and had purchased shares in several Cuban railroads. Further-more, with Spain dislodged from the Caribbean, the last real political obstacle to the eventual building of the Panama Canal had been removed.

Flagler was confident that with Spain no longer there to meddle, trade with Cuba and other Caribbean nations would soon explode, and that a deep-water port in Key West would not only be its logical staging point, but would also become a necessary stopping-off place for the enormous amount of ship-ping traffic plying the new canal route.

In contrast to his other terminal destinations, Key West was no undiscovered hamlet. With more than twenty thousand res-idents, it was the largest city in Florida at the time, and had been for more than fifty years. In the 1850s, U.S. Senator Stephen S. Mallory had dubbed it "America's Gibraltar," refer-ring to its strategic location at the mouth of the Gulf of Mex-ico, and the island city had a well-developed economy based on its status as a naval port and fueling station, as well as upon cigar-making, fishing, sponge diving, shipbuilding, and salvage.

Even if the practical issues involved in linking the island to the mainland by rail were daunting, the benefits to be gained by actually doing so seemed clear-cut. Flagler would become a dominant presence in what was then a major commercial cen-ter. His new rail project would serve not merely to connect one pleasure palace to the next, but to forge economic links between the United States and virtually all other nations.

Some have theorized that Flagler was motivated in this final undertaking by an underlying feeling of inferiority to his for-mer partner, John D. Rockefeller. Because the latter had con-tinued his close association with the workings of the company the two had founded, Standard Oil came to seem more Rockefeller than Flagler in origin. Others have theorized that all of Flagler's Florida undertakings were the result of a desire to "show off" for his new young bride, Ida Alice, even though most of what he accomplished there occurred after she had begun to lose touch with reality.

Still others speculate that it was the need to impress a third Flagler spouse, one Mary Lily Kenan, a woman he had first met in 1891, at one of the many social occasions arranged by Ida Alice during the early, halcyon days in St. Augustine. As Ida Alice's condition worsened over the course of the decade, it did not escape the attention of those close to the situation that one of the world's wealthiest men was spending an unusual amount of time in the company of his niece, Elizabeth Ashley, and her purported traveling companion, Mary Lily.

By the time Ida Alice was institutionalized for good, in 1897, the affair was a matter of speculation in gossip columns of the day, and Flagler, feeling pressure from Mary Lily's family as well as from the public, decided to make an honest woman of her. There was one obstacle, however: Flagler was still a married man.

In typical fashion, Flagler went immediately to work on the problem. In 1899, three weeks after proposing to Mary Lily, Flagler announced that he was moving his legal residence from New York to Florida. He allowed another two months to pass, and then petitioned the Supreme Court of New York that Ida Alice Flagler be certified insane and thus incompetent, a matter that could scarcely be contested, as she had for more than two years been locked in a private asylum, carrying on a one-sided conversation with the tsar of Russia and other imaginary suitors.

New York's divorce law was similar to that of Florida, however, in that divorce could only be granted where adultery could be proven. While Ida Alice had stated that she had indeed committed adultery on several occasions with the tsar, it was not the sort of contention that would hold up in court. So Flagler turned to more practical methods.

It took considerable doing, but on April 9, 1901, nearly two years after he had proposed to Mary Lily, a bill was introduced into the Florida legislature "to be entitled an act making incurable insanity a ground for divorce." Before the month was out, the bill had sailed through both houses and had been signed into law by the governor. Florida newspapers had a field day at Flagler's expense: it was rumored that it had cost him twenty thousand dollars in bribes to see the bill passed, and a number of "gifts" made by Flagler to Florida's public universities were documented.

Undeterred, Flagler filed for divorce from Ida Alice in June of that year, a petition that was granted in Miami, in August. During the proceedings it was revealed that Flagler had placed some $2 million in stocks and securities in trust for Ida Alice, providing her with an income of approximately $120,000 a year. As the costs of her care in what was reported as a "rich person's asylum" came to about twenty thousand dollars annually, some of the criticism of Flagler abated.

In 1988, biographer Edward N. Akin, in the first edition of his book *Flagler*, wrote that he had discovered notations in Flagler's papers documenting a payment to Florida legislator George P. Raney of more than $100,000, a sum written off as "expenses." It was a finding that seemed to lend credence to the notion that Flagler had paid enormous bribes to have his way.

In a later edition of his book, however, Akin admitted that he had overlooked a decimal point in the entry. The actual amount paid to the legislator, an attorney who often did work for Flagler, was slightly in excess of *one thousand dollars,* and in truth seemed to be reimbursement for various undertakings unrelated to the divorce bill. Akin added that while Flagler was entirely capable of bribery and had indeed made payoffs during his Standard Oil days, one thing was certain: he would never have thought it necessary to spend so much to sway a state legislator.

In any case, seven days after his filing, Flagler's engagement to Mary Lily was announced in newspapers throughout the South. Flagler was seventy-one, Mary Lily Kenan was thirty-four, and the public and press reacted accordingly.

Flagler, however, was delighted with his good fortune. He and Mary Lily were married on August 24, with Flagler joined this time to a woman who was by all accounts his social and intellectual equal, and who would prove a faithful companion through the rest of his days.

Flagler spared no expense to make his new wife happy. As she had once remarked that she had always wanted "a marble palace," Flagler built one for her: a mansion overlooking Lake Worth, which they named Whitehall, and which was to become a fabled center of the Palm Beach social scene. Among the many who joined the Flaglers' gatherings there were Admiral George Dewey, John Astor, Henry Sloane, and Flagler's longtime friend Elihu Root, secretary of war under Theodore Roosevelt.

It was Root, significantly enough, who had encouraged Flagler's interest in Cuba and the Panama Canal early on. Root was convinced that U.S. interests in the Caribbean were about to undergo a dramatic upswing and that Flagler, with control of the sole rail link to the nearest U.S. port, would profit greatly.

Speculation about Flagler's intentions had been rife for

more than a dozen years. It had been reported by one Florida paper as early as 1895 that he had sent land agents into the Keys, where he had purchased as much as 50 percent of the northernmost island of Key Largo, using assumed names to avoid driving up prices.

Whether or not that is true, it is well documented that Flagler planned his actions carefully. Though he might appear to have acted in haste at times, as with his controversial divorce from Ida Alice and his marriage to Mary Lily, Flagler's actions were generally undertaken as the culmination of a meticulous process of preparation on his part.

Thus, while some were quick to label his subsequent announcement—the decision to extend the Florida East Coast Railway to Key West—as capricious, ill-planned folly, nothing could have been further from the truth.

8

The Eighth Wonder
of the World

Flagler's official announcement that he intended to
ride his own "iron" across the Straits of Florida to Key West
did not come until July of 1905, but it seems clear that he had
been destined to make the attempt from the midsummer of
1898, at the conclusion of the Spanish-American War. From
that point forward, FEC documentation suggests that Flagler
and his top-level managers had begun to study the proposal in
a new light, going so far as to commission a number of preliminary engineering reports and feasibility surveys. The "railroad across the ocean" had at long last shed its pie-in-the-sky
status, and was now being taken seriously by one of the most
successful entrepreneurs in America.

Characteristically, Flagler attempted to keep his plans close
to the vest, in order to avoid driving up the price of right-of-
way acquisition, among other things. Though previously he
had had the luxury of choosing a number of alternative routes
in considering the exact location of his railroad-building proj-

ects, his options in the case of the Key West Extension were not nearly so broad. And experience had taught him what obstacles the avarice of local speculators might create.

In the 1890s, when Flagler had announced his intention to extend his line along the central coast to Palm Beach, a group of landowners from the then-prosperous town of Juno had formed a consortium, pooling their holdings in order to force up the price Flagler would have to pay for right-of-way. The group was certain that Flagler would have to meet their inflated price in order to avoid sending the line through a broad swath of marsh that would drive the cost of construction through the roof.

Presented with the consortium's demands, Flagler ordered his railroad built exactly where his foes had assumed it could not be done. At tremendous expense, Flagler's railroad went southward to Palm Beach, nonetheless, and the town of Juno, then the county seat of all South Florida, withered and died.

With this experience in mind, a glance at the topography of the Florida Keys made it clear to Flagler that the possibilities for locating his line were, literally, quite slender.

The Keys had been described by one writer of the day as "worthless, chaotic fragments of coral reef, limestone, and mangrove swamps . . . and have been aptly called the sweepings and debris which the Creator hurled out to sea after he had finished shaping Florida."

Key Largo, northernmost of the Keys and nearly forty miles from tip to tip, is by far the largest in the chain, but even that island is a few hundred yards wide or less in many places. Farther south, the Keys shrink dramatically in size, some mere dots of "land" the width of a football field in spots, many of them separated by miles of open water.

As a consequence, Flagler spent a great deal of time and

money investigating just what route his "impossible" railroad might take. And while he awaited word from his project managers, there had been plenty to attend to during the period from 1898 to 1905.

Though his activity in the company had diminished significantly, he was still a member of the board of directors of Standard Oil, the largest corporation in the world. And as the chief executive officer of the Florida East Coast Railway and its various subsidiaries, he was called upon to oversee a vast network of undertakings that stretched the entire length of Florida, including extensive freight and passenger operations, the management of a wide variety of hotels and resorts, the direction of a massive land sales and development operation, and much more. He had built a hotel in Nassau, for instance, calling it the Colonial, and had formed a shipping line—the Peninsular and Occidental Steamship Company—to service his offshore holdings, as well as provide freight and passenger service to Cuba and elsewhere in the Caribbean.

And, as usual, there was always the unexpected to deal with. In Miami, where Flagler's three-year-old city was finally gathering steam, a yellow fever epidemic broke out in the fall of 1899, spread by infected ticks brought ashore on a cattle boat from Cuba. Hundreds fell ill and the entire city was quarantined, with all boat and train traffic halted immediately.

Once again Flagler was pressed into service as savior of his "domain." Since the local economy was utterly dependent upon tourism and trade, Flagler poured his own money into an extensive program of public works, giving employment to a sizable labor force engaged in road and sidewalk building, port improvements, and the expansion of health-care facilities. By the spring of 1900 the epidemic had been controlled, and the quarantine lifted. Miami was on the road to a startling recovery: nearly written off by some observers at the beginning of the century, by 1910 "the city that Flagler built" would be hosting more than 125,000 visitors during its tourist season.

Meanwhile, with Spain removed from the equation, intricate U.S. diplomatic negotiations concerning the location of a canal linking the Caribbean and the Pacific had been moving forward, slowly but surely. In February of 1904 the United States Senate voted 66–14 to ratify the Hay-Bunau-Varilla Treaty, putting to an end the nearly thirty-five years of controversy and uncertainty surrounding the issue of where to build a canal joining East with West.

The bill secured the agreement of the newly formed government of Panama that the canal would be built across Panamanian soil by the United States, which, in exchange for a payment of $10 million, would hold virtual sovereignty over the ten-mile-wide "canal zone" in perpetuity.

The agreement also provided for a payment for French holdings in Panama amounting to $40 million, which was the largest real-estate transaction in all history. The fact that the government had paid more for the rights to build and control the Panama Canal than had been expended on the Louisiana Purchase, the Alaskan territory, and the Philippine Islands combined is some measure of the excitement generated by the culmination of the decades-long effort to forge an intracontinental connection between the world's great oceans.

Adding to the intrigue was the fact that Theodore Roosevelt had secured such favorable terms for the deal only by agreeing to back Panama's secession from Colombia. It was an arrangement typical of the egocentric Roosevelt, but one that did not sit well with his own cabinet.

"I took the Isthmus," President Theodore Roosevelt was to declare often, pointing to the treaty as his most important action in office. It may have been Roosevelt's most profound achievement, but it was scarcely without controversy.

As Elihu Root, Roosevelt's secretary of war, observed, "You certainly have, Mr. President. You have shown that you were accused of seduction and you have conclusively proved that you were guilty of rape."

In any event, the Hay treaty gave Flagler the practical justification he needed to go forward. With the canal's future assured, it seemed that Panama would become, in Simon Bolívar's words, "the emporium of the universe," and Henry Flagler would be poised to take advantage of it. He would build his railroad to Key West, where his planned deep-water port and storage facilities would replace Tampa as the closest deep-water port to the new canal, by more than three hundred miles.

In addition to servicing the veritable flood of shipping to and from the canal, Flagler contended that his new railroad would also provide a necessary link in an increased traffic—tourist and trade alike—with the liberated nation of Cuba. As it turns out, Flagler had met with Canadian railroad magnate Sir William Van Horne, and had come away impressed with Van Horne's plans to deploy a wide-ranging network of freight lines to service the island's pineapple-, sugar-, and tobacco-growing regions.

Furthermore, as the only naval station located on the Gulf and southern Atlantic coasts, and a mere ten miles distant from the main shipping channels, Key West was also ideally located as a coaling station for Navy ships patrolling the strategic waters of the Caribbean. In short, Flagler argued, Key West would become the foremost port on the Atlantic coast of the United States.

Whether Flagler believed his own rhetoric is debatable. By

1904 the Florida East Coast Railway had yet to turn a significant profit on any of its operations, with losses running anywhere from $100,000 to $400,000 in a given year. His string
of hotels showed erratic returns, just about breaking even at
best.

Though company records are often murky on the issue, it is
certain that had Flagler been content simply to plow his dividends back into Standard Oil stock, his fortunes would have
prospered exponentially. According to figures in his own sta-
tistical diaries, with upwards of thirty thousand shares in
his name, Flagler's income from dividends was as much as
$150,000 a *month* during the first part of the decade.

But by this time Flagler was seventy-four years old, and was
accustomed to his place as savior, if not outright ruler, of his
domain. In his second career, he had inarguably left behind his
status as an accumulator to become a builder, certainly the
most prominent developer in the Florida frontier, and, according to some, the man who singlehandedly created it as a modern state.

He had husbanded Florida's economy through at least two
natural disasters, and had established every one of the glittering, world-famous beach resorts that ran the length of its eastern shores. By the time he had reached Miami, some might
have assumed Flagler was at a natural stopping point.

He could retire at long last, one might have conjectured,
there to join his new bride and—by all accounts—boon companion, Mary Lily, at their fabulous home in Palm Beach,
where he could rest and enjoy the fruits of his storied labors.

Instead, he continued on.

That Flagler chose the latter path says more about the man
than any other action undertaken in his lifetime. In choosing

to build his railroad to Key West, Flagler was tackling a project that in some ways outdid the prospect of the Panama Canal. For at least in that instance there was general accord that such a giant ditch could indeed be dug, if one were to expend sufficient energy and manpower and money and materiel to see the job through. In the case of the railroad across the sea, however, few were as sanguine as that.

9

Charting the Territories

Company records indicate that since reaching Miami in 1896, Flagler had been making tentative moves southward, feinting here and there like a fighter waiting for the right time to wade in for real.

In 1903 he had added a twelve-mile extension of the FEC along an ancient limestone formation running southwestward from Miami and known as the Cutler Ridge. At scarcely ten feet above sea level, the "ridge" was more of a barely perceptible rise, dotted with outcroppings of palmetto scrub and slash pine. But its elevation made enough of a difference, agriculturally speaking, to allow for flourishing fruit and vegetable crops in the area south of the growing city, and Flagler took quick advantage of this by adding freight service to the area.

In 1904 Flagler extended the Cutler line another sixteen miles southward, establishing a tiny settlement in a place known as Homestead, which constituted, to all intents and purposes, the last reliably dry land on the continent. And it was here that Flagler's line languished, while he pondered the final decision on which direction his railroad would take to Key West.

The options were essentially two. The first was to go straight south from the Homestead terminus, building roadbed a few miles out along a narrow isthmus, and then jump by bridge across narrow Jewfish Creek to Key Largo, at a point about a quarter of the way south along that island's span. From there the route would follow the line of the Keys, leapfrogging where necessary across the open waters that separated them, though in some cases, Flagler theorized, dredging operations could actually succeed in joining the islands that were closest together, making two keys into one where feasible.

The other choice involved building his road southwestward from Homestead, across the very tip of the Florida mainland, to an outpost known as Cape Sable, where the road would leave the continent and strike out over forty miles of uninterrupted open water to join land somewhere in the Lower Keys. To investigate this route, Flagler dispatched a civil engineer named William J. Krome, whose name still graces the westernmost avenue in far-flung Dade County.

Krome and his surveying party were to encounter the same daunting conditions that James Ingraham and his men met while crossing the Everglades in the opposite direction a decade before: endless stretches of marshland and muck, dense stands of ten-foot-high saw grass with edges as sharp as razors, clouds of stinging insects so thick "you could swing a pint can about on the end of a string and come up with a quart of mosquitoes."

Modern-day travelers can get a hint of what Krome was up against by following the forty-mile road that winds from just outside Homestead to Flamingo, near where Flagler hoped to send his route. The route is circuitous, following a twisting "ridgeline" through veldtlike pastures interspersed with marshlands alive with shorebirds and alligators and the ever-present mosquitoes, which descend in literal clouds upon the unwary.

A distance that can be traveled by car in an hour or so today took Krome and his men thirteen days. "I found a most God-forsaken region," he wrote in a report to his supervisor. "But of keys, bays, rivers and lagoons there is no end, and it is going to take us much longer to get a survey than I had expected."

Krome did press on, often forced to drag the shallow-bottomed boats they had brought along over terrain that was an indefinable mix of peat and muck and water, sucking at every footstep, yet not liquid enough to cross by boat. The men were tortured by heat, humidity, and insects, and often lost their way in the featureless landscape. If not for the aid of the occasional backcountry hunter or member of the native Miccosukee tribe, Krome's only memorial might have been a long-forgotten pile of bones.

But Krome managed to find his way to open water, and lived to warn his employers against any notion of sending a railroad to follow in his footsteps. Though he did conclude that a route might be feasible down the eastern edge of the cape, the costs would still have been excessive, especially when the jumping-off point would have been situated much farther away from the Lower Keys than Flagler desired.

According to Carlton Corliss, an engineer who worked on the project for many years, Flagler also considered the possibility of bridging a set of narrows across Baines Sound, near the present-day Card Sound Bridge. The extension would go no farther than the northern tip of Key Largo, where a deep-water port would be dredged and docks constructed at Turtle Harbor. That plan had the distinct disadvantage of leaving Key West out of the equation, however.

In the end, both the Turtle Harbor and Cape Sable routes were rejected, and Flagler turned his attention to the route down the Keys, the path that Jefferson Browne had outlined in his *National Geographic* article of a decade before. According to Browne, such a route had actually been surveyed as early as 1866, by a civil engineer named J. C. Bailey, who had been employed by the International Ocean Telegraph Company to chart the course of a telegraph line from Miami to Cuba. To Bailey it seemed a natural adjunct to lay a rail line along the same path.

It might have seemed logical to Browne and Bailey, but a glance at the actual terrain to be crossed makes it clear why Flagler hesitated before choosing. Of the more than one hundred miles that remained from Jewfish Creek to Key West, about half were over open water. Some of the spans were a few hundred yards. Others stretched for miles.

Still, to a pragmatist like Flagler, the route seemed possible. When questioned how he would cross those mammoth stretches of open water, Flagler replied, "It is perfectly simple. All you have to do is to build one concrete arch, and then another, and pretty soon you will find yourself in Key West."

In any case, by January of 1905, Flagler felt that there had been enough of data gathering and surveying. He called his general manager, Joseph R. Parrott, into his office and announced that it was time a decision be made.

"Joe, are you sure this railroad can be built?" Flagler is said to have asked.

Parrott, whose job it had been to evaluate the endless series of reports and surveys, said, "Yes, I am."

"Very well, then," Flagler said. "Go to Key West."

Late that month, Flagler traveled to Key West along with Parrott and other FEC executives and announced to a group of the town's leading citizens that he would indeed bring his railroad to their city. Back in Tallahassee, Senator E. C. Crill of Palatka introduced S.B. 11, granting the Florida East Coast Railway rights and privileges to build the Key West Extension and granting the company a two-hundred-foot right-of-way down the Keys. The bill became law on May 3, 1905.

In the ensuing months, newspapers from Florida to New York were abuzz with details of Flagler's plans. Not only would the FEC build the railroad extension down the Keys, it would also construct twelve piers in Key West to handle the expected shipping traffic. Each pier would be two hundred feet wide, with covered storage, and would extend seaward eight hundred feet, flanked by docking basins two hundred feet wide, allowing berthing space for four oceangoing freighters or passenger ships. All this was to be completed, Flagler asserted, by January 1, 1908, or scarcely two and a half years later.

Meanwhile, Flagler had placed advertisements in a series of newspapers nationwide, soliciting private bids for construction of the railroad itself, and he was taken aback to find that only one proposal had come in, and that was a cost-plus bid, wherein the contractor would be guaranteed reimbursement for whatever expenses were incurred, plus a previously agreed-upon amount for profit. Flagler balked at such a deal, of course. All his life he had taken risks, and the idea that anyone should shoulder none himself, while expecting a handsome return, was simply unacceptable.

He and Parrott thus determined that the job would be undertaken in-house, and they began a search for a likely project engineer. Flagler was confident that he had plenty of men on staff who were capable of building roadbed and laying track on dry land. He'd been at such projects for two decades now. But since half of the Key West Extension involved laying track over water, it was obvious he would need a master bridge builder to oversee this project.

Eventually the search led them to Tampico, Mexico, where a young engineer, Joseph C. Meredith, had spent two years in charge of a massive $3.5-million, half-mile-long pier construction project undertaken by the Missouri Valley Bridge Company for the Mexican government.

Ralph Paine, who wrote a lengthy piece on the project for *Everybody's,* a popular magazine of the day, described Meredith as "a taciturn, almost diffident person," but he was also an expert in the use of reinforced concrete, a construction medium Flagler had first been intrigued by when building the Alcazar in St. Augustine.

Bridging some of the endless stretches involved in this project using conventional ironwork was out of the question, but as Flagler had suggested in his earlier comments, a series of poured columns and arches, built in the manner of the ancient Roman viaducts, seemed feasible. And Meredith, who might have been "fragile-looking," but nonetheless had the reputation for taking on projects that seemed too big or problem-filled for others, seemed the perfect person for the job.

In an interview, Meredith remembered being summoned directly from Mexico to St. Augustine for his interview:

They told me about the Key West extension. Not a word about cost or possible profits; merely the matter of engineering feasibility. Mr. Flagler wanted either to fill in or to build a viaduct, for he hates makeshifts. Permanence appeals to him more strongly than to any other man I ever met. He has often told me that he does not wish to keep on spending money for maintenance of way, but to build for all time. . . . [A] corporation, especially where the country has to grow up and the paying traffic is all in the future, will do barely enough to supply the pressing needs. They make improvements gradually, as the profit comes in. But that is not Mr. Flagler's way.

Their interview was brief and to the point. Once Meredith had assured Flagler that nothing about his plans seemed impossible, the matter was settled.

"When can you start?" Flagler asked Meredith, fully expecting that he might ask for a month or so to settle his affairs.

"I'm ready to go to work this afternoon," Meredith told him. "But I'd like a few days first to go home to Kansas City, pack some things, and see my family. I'll have to be on this job for several years."

"All right, my boy, see your family," Flagler told him, and then echoed what would become a familiar refrain: "Then go to Key West."

10

Jumping-Off Point

There were many who applauded Flagler's announcement, of course. As former secretary of war Elihu Root wrote: "I regard it [the Key West Extension] as second only to the Panama Canal in its political and commercial importance to the United States."

But when Flagler confided to his old friend George M. Ward, a Presbyterian minister and president of Rollins College, that he had hired a supervising engineer and embarked upon the project at long last, Ward's reply was less than encouraging. "Flagler, you need a guardian," Ward said, a sentiment regarding his sanity that would be shared by many others of the day.

"Flagler's Folly" was the epithet most fondly used by Flagler's detractors, and evidence suggests that even Flagler himself was not always so sure of his intent. Flagler biographer Edward N. Akin notes that in a version of his will penned in 1901, the great man expressly enjoined the use of his estate for

the building of any railroad extension south of Miami, a provision he would ultimately be required to obviate.

But by 1905, Flagler had made his decision, and no amount
of public derision was about to sway his course. Company-
generated public-relations material of the time resounded with
its founder's resolve and sense of greater purpose:

> As has been intimated, the assurance of the Panama
> Canal made the world look at the Keys of Florida
> and Key West from a new point of view. The canal
> opens in a moment tremendous vistas and pushes
> our commercial horizon across the seven seas. Key
> West is almost three hundred miles nearer the east-
> ern terminus of the canal than any other of our Gulf
> ports. At the same time it is the natural base for
> guarding and protecting the canal on the east and
> our great Gulf coast. That the island should be
> closer to the mainland had been the dream of gen-
> erations. The dream had become a necessity to our
> commerce, our national interest, and our national
> safety. But could the dream come true, could the
> necessity be met?
>
> The financiers considered the project and said,
> Unthinkable. The railway managers studied it and
> said, Impracticable. The engineers pondered the
> problems it presented and from all came the one
> verdict, Impossible. . . .
>
> But, strange as it may seem, there was a financier
> with the courage of Columbus, a railway manager
> with the administrative grip of a Menéndez, and an
> engineer as brave and as far-seeing as the pilots who

brought the caravels of Spain through miles of
unknown and uncharted seas. . . .

Thus did Flagler, apparently characterized as Christopher
Columbus, undertake his "folly" aided by Parrott, his Menén-
dez, and his Cortez, Joseph C. Meredith. Overheated rhetoric
aside, the project was an undeniably grandiose one. Some writers noted that the route was longer than that of the hundred-
mile-long Suez Canal, which had taken ten years to build. Just
as remarkable was the announcement that all work on the
project would be carried out by the FEC itself—nothing would
be contracted to others.

And the costs were unprecedented as well: the 742-mile California link of the Union Pacific Railroad, joined by the driving of the "golden spike" at Promontory Point, Utah, in 1869,
had cost $23 million. Total budget for the Overseas Rail-
road—estimated by some analysts of the day at $50 million or
more—would soar well beyond $27 million.

The sum had to seem formidable to Flagler himself. All his
years of doing and building in the state barely totaled as much.
The best estimates suggest that by 1905, Flagler had invested
about $30 million in his Florida enterprises, $12 million in
hotel building, and another $18 million in railroad building.
Much of the cash for the investments had been raised from
loans issued against his Standard Oil Stock. In turn, the various business units listed Flagler as the prime creditor, paying
off their debts to him out of revenues. In turn the business
units, as well as the vast landholdings that came along with
right-of-way development, were often used as collateral for
further development.

To begin such a task as this, then, at the age of seventy-five, suggests that the man could not have been concerned primarily about taking his fortunes with him. He had overridden the advice of a number of financial analysts within his own company, after all, and had firsthand evidence that no other builder in the world was willing to share in the challenges that lay ahead.

Henry Flagler, it seems, had grown far beyond his place as a mere businessman, at least in his own mind. In a letter written to a clergyman friend, he described his feelings for the citizens of Florida in a telling manner: "I feel that these people are wards of mine and have a special claim upon me."

If the letter is suggestive of a benevolent despot's self-aggrandizement, Flagler did not strike most observers that way. Even at his age, Flagler had a bearing that would make it difficult to imagine anyone calling him crazy to his face. Edwin Lefevre, then a leading financial writer for the *New York Sun,* was sent to interview Flagler for a story in *Everybody's,* and described him this way:

> *His hair is of a clean, glistening silver, like the cropped mustache and the eyebrows. They set off his complexion, which is neither ruddy nor baby-pink, but what one might call a virile red. He has a straight nose and a strong chin. . . . The eyes are a clear blue—some might say violet. They must have been very keen once; today their expression is not easy to describe—not exactly shrewd nor compelling nor suspicious; though you feel they might have been all of these, years ago . . . eyes that gleam but never flame. . . . A handsome old man! Under his fourscore years his shoulders have bowed slightly but there is no semblance of decay.*

Lefevre, who embarked upon his assignment ready to deliver a portrait of a robber baron facing his just, fin-de-siècle desserts, was to spend several weeks inspecting Flagler's vast holdings and several more days in conversation with the man; he came away with a vastly revised assessment. Toward the end of his series of interviews with Flagler, Lefevre reflects:

> You realize that you are before a man who has suf-fered and has never wept; who has undergone intense pain and has never sobbed; who has never bent under stress and has never hurrahed! . . . Your great man is apt to be one with certain faculties over-developed, and classifies easily. But Flagler is not like any one else and withal is not eccentric.
>
> He is without redeeming vices, without amiable inconsistencies, without obsessions. He simply does not "classify." You cannot accurately adjectivize him. He does not defy analysis; he baffles it. . . . Whether [his veins] run red blood you cannot tell; but you are certain it is not ice water. What color is it, then? That is the mystery of the soul of Henry M. Flagler.

Such speculation may have intrigued others, but Flagler showed little awareness of any mystique surrounding him. He might have been a rich and powerful man, but his own preoc-cupations seemed to center on what he might accomplish of a day. His old friend Reverend George Ward would recall that Flagler rarely read history, for he was interested primarily in the present and, most of all, the future.

In a letter written to an associate at the time he was about to announce the undertaking of the Key West Extension, Flag-

ler provides a glimpse of his own self-image at the age of
seventy-four:

> *I was born with a live oak constitution, and it is only
> within a year or two that I have known of the pos-
> session of any organs. My diet has always been sim-
> ple and the only excess I believe I have indulged in
> has been that of hard work. I have however one ail-
> ment (old age) which is incurable, and that I am sub-
> mitting to as gracefully as possible. I am quite sure,
> however, that I possess as much vitality and can do
> as much work as the average man of forty-five.*

Hard work, energy, and accomplishment. For Flagler it
seemed to be all he knew, all he need know.

If Flagler's innermost thoughts remained an enigma to his
contemporaries, certain details of his plans for the work that
lay ahead were perhaps more revealing of his singleminded
determination. As Meredith heard during the first meeting of
the two, Flagler saw no reason why dredging operations could
not be used to lighten their task. The waters surrounding the
shoal-like Keys were shallow in most places, and it would be a
far simpler, less expensive proposition to simply dig up the surrounding sea bottom and use the material to fill in the gaps
between islands than to build bridges.

Initial surveys suggested that only six miles of typical bridge
spans would be needed along the more than 100-mile route.
The rest of the fifty-plus miles of open water could be filled in,
or traversed by the series of arches that Flagler envisioned. In
fact, one of the preliminary studies suggested that the entire
route could be constructed atop a solid rampart that could

wind its way down the line of the Keys like a version of the Great Wall of China.

The environmental issues raised by such an approach would today stop a project dead in its tracks. And even at the beginning of the century, government scientists were concerned about the potential for disrupting the natural tidal flow between the Gulf of Mexico on the west of the Keys and the Atlantic Ocean on the east.

Some went so far as to predict that the planned impediments, even with arched viaducts replacing the solid ramparts, would affect the character of the nearby Gulf Stream as it surged up from Cuba, thereby changing the climate of the Keys and all South Florida. In the end, Flagler's submitted plan called for six miles of open-water bridges, and upon that note of compromise, construction would begin. Jefferson Browne's decade-old declaration that "a railroad to Key West will assuredly be built" had finally been proven correct.

In April of 1905, a contingent of several hundred laborers, most of them Southern blacks, began the process of building up a roadbed across the last miles of marshy land south of Homestead. While there was no open water to cross, the terrain was typical of the Everglades: soggy at best, too wet to be considered land and not wet enough to be called water. Mule-drawn carts were useless, as was conventional motor-driven equipment.

To cope, Flagler's engineers devised a pair of shallow-bottomed dredges, each with a steam-powered excavating shovel mounted on its deck. Once the tangle of mangroves and brush had been cleared, the machines advanced on either side of the roadbed-to-be. Each scooped out the muck in front of it, thereby creating a canal for the machine's continued passage. The muck and limestone that had been removed were deposited in between the two roaring behemoths, to be graded into roadbed by the trailing laborers.

It was an ingenious solution, one of a series that would be required over the years to come, but the problems were just beginning. The combination of heat, humidity, insects, and isolation had its effects upon the workmen. It was, after all, no accident that the terrain Flagler's road was crossing had remained virtually uninhabited over the course of all known history.

One worker recalled that the dredge operators would return to their machines in the morning to find that alligators had occupied the decks during the night. The creatures had to be driven off into the swamps before work could begin again.

The simple fact was that no one had ever attempted to perform the arduous labor that railway or any other major construction required under conditions such as these. Exhaustion, heatstroke, and disease would soon take its toll, even among a workforce long accustomed to harsh conditions.

By the time Flagler and a team of FEC executives made an inspection tour in late July of 1905, the roadway had scarcely made its way to Key Largo—still 110 miles from Key West—and the original contingent of more than 400 workers had dwindled to fewer than 150. Flagler had made the trip to the construction site aboard a paddle steamer from Miami, the party accompanied by Russell Smith, an engineer researching a story for the *New York Herald*.

Smith's account began with a lyrical description of the

waters just off Key Largo—"Going over shoals and through such narrow channels that the boat touches both sides at once, the bottom of the sea is always visible, with its jelly-like sponges and beautiful branching coral. . . . 'Skipjacks' continually jumped out of the water and rode on the surface for hundreds of yards. . . . Flying fish left the water and flew for a quarter of a mile before returning"—a passage virtually as accurate today as it was a hundred years ago in evoking the charms of the nearby open waters.

But soon enough Smith's view of things changed. No sooner had the *Biscayne,* Flagler's paddle-wheel steamer, tied off at the houseboat serving as field headquarters for the project than the insects descended.

"Here is experienced the most serious problem to overcome this great undertaking," Smith wrote. "The mosquitoes on this key are almost unbearable, and the problem is to persuade laborers not to run away, for it means certain death as there is no possible outlet to the mainland."

The plaguing insects did drive the inspection team away, forcing the *Biscayne* to anchor a mile and a half offshore, where the captain assumed they would find relief. But as anyone who has spent a summer's night anchored within cannon-shot of land in the Florida Keys might have told the party, such was not the case. Once they'd had a taste of blood (or, more accurately, the human exhalations of carbon dioxide), the mosquitoes simply followed the party over the water like tiny flying bloodhounds.

Smith writes that no one on board the *Biscayne* was to get much sleep that night, least of all himself: "I lay down on a canvas cot with a double thickness of blanket, thinking this too thick for the pests to bite through. . . . I reckoned wrong, for they attacked from below and bit through the canvas cot."

The trip made a profound impact upon Flagler, who understood that a steady supply of labor was crucial to the success of his undertaking. Labor shortages, which had always plagued his road building efforts in sparsely populated Florida, would be greatly exacerbated here. As one manager would lament, "One of our most trying problems has been to take a big body of low grade men, take care of them, and build them into a capacity for performing high class work."

Every mile that the road pushed southward from Miami was another step from civilization. And if the principal workforce, most of them men descended from slaves, were finding the going on this project tough, it did not bode well.

From the outset, then, Flagler stressed to his managers that living conditions should be made as comfortable as possible. To lessen the difficulty of finding suitable campsites, large barges were dispatched from Miami to serve as portable work camps, some of them floating two- and three-story wooden dormitories that could be towed down the coast of the Keys as the line moved ahead.

Field kitchens were also stationed on such barges, to which "food of only the best quality . . . in abundance" was ferried three times a week and more. "Camps are clean," wrote a reporter for the *Chicago Daily News,* "food good, pure ice and water supplied to each camp, no liquor sold in or near."

Smudge pots, burning a mixture of kerosene and oil, cast a pall of largely ineffectual smoke over all the work sites and common areas, and workers who wanted them were issued head nets, though the description of the latter, as given by William Krome, by that time Meredith's assistant chief engineer, provides a vivid picture of the struggles the men had to undergo where even the most basic issues were involved: "[The best] was one adapted from Cape Sable squatters [suggesting that Krome's death-defying trek through that wilderness had provided at least one positive]. It is built for use over a stiff rimmed hat and consists of a band of canvas fitting closely around the crown of the hat. To this is sewed a strip of close mesh copper wire netting extending down the back and curving over the shoulders to the level of the chin. Cheesecloth is taped around the bottom of the copper gauze and tucked beneath the coat, which is buttoned over it."

While there is no reason to doubt Krome's claims of the item's relative effectiveness, it is another thing to imagine trudging through ninety-degree heat and equal humidity, swinging a twenty-pound sledge or manhandling a precarious wheelbarrow full of sloshing marl, with such a getup around one's head.

Still, one worker recalled that most of the men did wear screened hats when they were working, as well as gloves and long-sleeved shirts. They covered their bodies in oil and carried palm-leaf switches to try and drive the insects away, he said. But mostly they stayed inside as much as possible. "We rarely went outside after five P.M.," he said.

Aside from the debilitating nature of the work itself, maintaining a steady supply of fresh water constituted another major problem. Other than cisterns used to store and collect rainwater, no supply of fresh water was to be found anywhere on the Keys, then or now. The company drilled test wells as deep as two thousand feet, but found nothing but salt water.

Finally a system was devised whereby fresh water was pumped out of the Everglades to large holding tanks near Homestead. Each day, two trains made up of flatcars carrying huge water tanks fashioned from cypress planks ferried the precious commodity down the line to supply workers and their machines. By the time the line was nearing its latter stages, water was being hauled well over one hundred miles to men surrounded by a sparkling blue ocean that might as well have been an endless stretch of desert sand.

Ever mindful of the yellow-fever outbreak that had quarantined his new city of Miami scant years earlier, Flagler was concerned that a similar outbreak might strike his Keys workforce. Every work camp had a pair of trained attendants on duty, and those workers sick enough to require a doctor's attention were transported by rail back up the line to Miami or by ship south to Key West, where the company maintained hospitals and provided medical services free of charge to its workmen.

Dehydration, influenza, and an occasional rattlesnake bite, along with the fractures and lacerations common to heavy construction projects, made up the bulk of the medical referrals. Despite Flagler's concerns, no plague or tropical fever ever swept through his camps.

Of far greater concern to the workforce than housing standards and quality of medical care were the rigors of the work itself, especially when coupled with the isolation they would undergo. Flagler, who had been drawn to Florida because of its paradisiacal nature, after all, did what he could to alleviate boredom in the camps. Once the project had extended to Marathon, a tennis club was erected for the amusement of the engineers and supervisors familiar with that game.

But for the rank and file, the types of diversion they were accustomed to in less far-flung territories proved hard to come by. Flagler, perhaps because of his conservative upbringing, was no drinker, but even if he hadn't been restrained in his own habits, his absolute prohibition against liquor in the camps would have made for sound labor-management practice.

Women, too, were in short supply. However, in the later stages of the project, an upper-echelon employee such as bridge-building expert and division engineer C. S. Coe might be able to quarter his family in company housing at one of the larger work camps, though such permanent arrangements were rare, and for a laboring man they were out of the question.

Flagler did permit periodic visits by family members of workmen, however, and as the camps developed, he provided limited accommodations for visitors. One of the "boarding camps" he built along the way, set aside for the enjoyment of the families of skilled workers on the gangs, later evolved into the Long Key Fishing Camp, which would ultimately prove a popular destination for Flagler's more affluent clientele.

Such diversions would remain out of the reach of the average worker, however, and it was not long before American ingenuity proved itself equal to the task of providing diversion for Flagler's minions. Enterprising local residents were often able to supplement a hardscrabble livelihood in farming or fishing by smuggling in "hootch" to thirsty workers, and one group of the more visionary went so far as to refit an ancient freighter and give it a new life as a floating saloon-cum-bordello to which a desperate man might catch a ride via skiff for an evening's entertainment.

Such excursions were anathema to Flagler and his top brass, of course, and district supervisors were under strict orders to keep the booze out of the camps and the men safely in their quarters at night, an ideal that might have made sense in the cool boardrooms of St. Augustine, but one hard to live up to for the manager of a far-flung labor camp where men were always on the verge of desertion, even if it meant risking one's life. Managers often looked the other way, then, when overloaded skiffs slipped away from the "quartersboats," headed toward the distant lights of a party ship, and even a number of the company stewards were not above supplementing their incomes by dealing in black-market whiskey.

One division engineer wrote to Meredith, pleading for some discretion. "I have found great difficulty in the prevention of the liquor traffic . . . ," he said, adding, "The worst that we have had to contend with . . . was the unfaithfulness of our own steward in the matter." Another difficulty he bemoaned was the fact that often men in a camp as strictly run as his own knew full well how easy liquor was to obtain in another camp up or down the line.

This supervisor went so far as to suggest that the railroad attempt to control the situation by establishing its own system of "canteens," but the proposal was dismissed by Flagler, who felt that it would be tantamount to official capitulation on the issue.

Another company supervisor, W. R. Hawkins, wrote in his diary of the periodic need to round up workers who had defected upon hearing stories of "oil finds" up and down the right-of-way. Crew superintendents investigating the claims would invariably find a group of excited workmen surrounding an upwelling of sulfurous swamp gas that had spread its oily residue over the nearby waters.

At any rate, the difficulties in finding and keeping laborers led to an ever-increasing set of recruitment efforts by company officials, who turned their sights on every likely source, including Bowery bums, Greek sponge divers, Italian steelworkers, and a considerable number of native workers from the Caribbean, who were more familiar with the climate and the terrain.

One of the most dependable groups turned out to be natives from the British Cayman Islands. The Caymaners would arrive at the work sites in their own boats early each January and work steadily until just before Christmas, with no complaining and few desertions. According to Benjamin Grinwell, a long-time Extension employee, the men were a mixture of British and West Indian, many with sandy complexions and red or blond hair, but with otherwise Negroid features, and "no more than three or four surnames among the hundred or more of them." Whatever their lineage, the Caymaners proved to be expert helmsmen and soon became the backbone of Flagler's varied fleet.

By the time crews had hacked their way the twenty miles or so to Jewfish Creek and began serious work down the spine of Key Largo, the workforce had grown significantly, averaging some three thousand men at a given time, reaching four thousand at the project's most fevered peak. If, at times, the camps resembled a staging area for the Tower of Babel, relations among the various groups remained largely placid, perhaps because the men were forced to unite against the difficulties of the work and the environment in which they lived.

Flagler, who had been called before his share of investigative panels in connection with the workings of Standard Oil, had been spared such scrutiny for the most part during his days in Florida. But with the eyes of the world turned upon the Key West Extension, things were sure to change.

One of the main controversies arose over the activities of labor recruiters who contracted independently with the FEC. These men, or *padrones,* as they were called in the Italian communities, did much of their work in urban centers of the Northeast, including New York and Philadelphia.

Such a contractor was technically an independent agent, an arrangement that critics said was set up to shield the company from any improprieties in the recruitment process. The contractor would promise to deliver a certain number of "able-bodied men" to FEC offices in Miami, in return for cash—sometimes one dollar a head, sometimes as much as three dollars—or in some instances, a grant of land from the vast right-of-way holdings which had been amassed.

The recruits, who hired on at rates that ranged widely according to their skill levels (skilled carpenters might get two to three dollars for a ten-hour day) and the degree of their desperation, came to Miami aboard FEC trains, with the understanding that the twelve-to-fifteen-dollar ticket charge would be deducted from their first month's check. In Miami they would be transferred to trains bound for the work camps themselves, where they would be housed either in one of the floating camps or in one of the land-based dormitories. Some of the larger structures might sleep as many as 350 men in a series of bunks four shelves high, though most of the buildings were far more modest, with forty or fifty workers housed in each. The floating quartersboats were somewhere in between in capacity, usually housing as many as 150 men.

While the circumstances the men were to encounter at the camps were daunting, exploitation of a sort worked both ways. One of the headaches for the recruiters and company officials alike was the proclivity of their recruits for disappearing along the way from New York or Philadelphia to Miami. One internal company memorandum complained that as many as half of the men recruited in New York during the early days of the project had vanished from the train before reaching the work camps.

In one of his diary entries, a company supervisor offers some insight into the problem in a description of a trainload of new recruits that reached Marathon: "Two hundred sixty-nine [of 345 that boarded the train in New York] mostly pretty tough looking customers, though most anyone would look tough after a trip from New York in day coaches with no chance to wash and not much chance to sleep nor eat."

Men took advantage of their deferred fares simply to walk off the trains in Jacksonville or Palm Beach or even Miami, there to seek work in more hospitable circumstances and avoid that twelve-dollar transportation charge. Others, claiming that they had been deceived by the trumped-up claims of the *padrones*—no mention of the transportation charges, their promised wages wildly inflated to as much as $1.75 a day (most common laborers actually received $1.50 for a day's work)—demanded immediate passage back north.

The claims and counterclaims began to attract attention, especially since the railroad's founder had been previously characterized as a robber baron by certain quarters of the press during the Standard Oil hearings. The result was a series of well-documented investigations by governmental agencies into claims of slave-labor conditions in the Flagler camps.

A special assistant U.S. attorney named Mary Quackenbos, who had been appointed to investigate similar labor-practice claims in agriculture and other industries in the frontier state, was quick to turn her sights on Flagler's recruitment operations. Attorney Quackenbos secured a deposition from one recruiter who claimed that the railroad had agreed to pay his firm three dollars apiece for laborers delivered to Florida. To recoup the costs of this extraordinary bounty, the company deducted two dollars from each of the new hires' pay, disguising it as part of transportation fees.

According to the deposition, the recruiters' efforts were centered on the most luckless of the available labor pool in New York City: the homeless, the charity wards, and ethnic relief agencies. The men, the government charged, were promised ideal working conditions, inflated rates of pay, and, in some cases, supervisory or skilled positions that never existed. Most were not told that the work took place on isolated islands from which the only way back was by company train, nor were they aware that return passage would be blocked until they had worked sufficient weeks or sometimes months in order to satisfy their debts to the company.

Stories circulated that some men did attempt to swim or slog their way through the swamps toward civilization, and that when caught they were forced to return to the labor gangs, this time as unpaid convict labor. Public outcry reached its peak in 1907, when a New York federal grand jury indicted the FEC and a prominent New York labor recruiter on the grounds that they had violated an 1866 law prohibiting the use of slave labor.

Because the allegations against the company and the recruiters had come in the main from witnesses whose reliability was open to challenge—many had criminal backgrounds, most were uneducated, others were alcoholic—the government abruptly dropped its case just as the matter was about to go to trial. But Flagler was far from pleased. To friends, he expressed his oft-repeated suspicions that the entire matter had been concocted by old foes of his in the government, still miffed that they hadn't been able to prove their charges against Standard Oil.

In one letter to John Sleicher dated November 20, 1908, he expressed relief that the case against the FEC for "Conspiracy, Peonage, and Holding Men in Slavery" had been dropped, but said, "There is however an air of mystery about this thing that we do not understand."

In another letter he wrote that same day to his old friend Dr. Andrew Anderson, Flagler was more pointed: "I have no idea that Teddy's [Roosevelt's] venom will permit him to stop as long as there is a chance to get a whack at me."

Flagler, who had been adamant from the project's outset about the need to treat his workforce well, also seemed upset that he had not gotten the chance to present a picture of actual conditions in the camps to the public at large. In a letter to Elihu Root, he reminded his longtime supporter that "when we started the work, I gave orders . . . that no pains or expense should be spared to house and feed the men in the best possible manner." He reiterated that the company had gone so far as to build hospitals in Miami and Long Key and keep them staffed with registered nurses, physicians, and surgeons.

Flagler also noted that Major General J. R. Brooke, another associate from the early days in St. Augustine, had often accompanied him on progress inspection trips down the grade, and that Brooke had remarked to Flagler on a number of occasions that he had never seen any U.S. troop installations where the men were as well quartered and fed.

From a more distant perspective, there seems little doubt that many outlandish claims were made by zealous recruiters (and embroidered by men yearning to hear tales of fountains of youth and streets paved with gold). But it seems equally clear that there were never slave-labor practices in effect on the project.

A telegram that Joseph Parrott, Flagler's operations manager, sent to Meredith in the midst of the furor sheds some light on the matter:

You will recall that more than a year ago I had occasion to advise you that some one had reported to me that some of our men on the extension work were using guns for making the men work and I told you at the time as a humane proposition it must not be done, and our men must be treated kindly, and if we had any foremen who could not work and supervise labor without the use of force or threats, or violent language they must be relieved. Apart from the humane idea, it is a business proposition that men work better and they are better men, if treated like men should be. While we may have some men among us who are not fit to be treated like men, I think it is best to discharge them even though it is an expense to us.

It takes no great leap, however, to imagine the dismay that must have been felt by significant numbers of men who were brought to a seeming paradise, there to encounter work of such a nature and difficulty as they could scarcely comprehend. It must have felt to some as though they had become slaves, captive to a project that would grind them to fishmeal unless they could somehow find escape.

11

A Surprise, the First of Many

A dependable source of labor was not the only problem Flagler faced, of course. In preparation for the project, he had amassed a formidable array of equipment, including three tugs, eight Mississippi River stern-wheelers, more than two dozen motor launches ranging from five to fifty horsepower, a dozen dredges, eight concrete-mixing machines to be installed on barges, two more concrete mixers for land, nine pile drivers for underwater supports and two more for land-based track pile driving, ten power shovels, a specialized catamaran for use in building forms for concrete piers, a pair of large steel barges and scores of smaller ones, several locomotive cranes, a floating machine shop, more than a dozen houseboats of varying sizes, and two oceangoing steamships.

Because much of the work would go on around the clock, all of the floating equipment was outfitted with its own generators to produce electric light. As word began to circulate about the scope of this assemblage, independent estimates of the costs of the project soared, with a reporter from the *Brooklyn Eagle* setting the ultimate price tag at more than $50 million.

It was far too late for any second-guessing on Flagler's part, however. In a process regularly chronicled by the *Palm Beach Daily News,* the *St. Augustine Record,* and other Florida and New York newspapers, six of the dredges had gone to work immediately on the nearly twenty-two miles of palmetto scrubland and mangrove swamp separating the Homestead terminus from Jewfish Creek, where the FEC line would cross to Key Largo.

The big machines were working from either end toward a meeting place midway between Homestead and the creek, their progress often delayed when the channels had to be hacked to a depth that would permit the dredges to proceed. As company officials would remark, it was a "web footed proposition from start to finish."

At the same time, crews were at work on a quarter-mile section of bridge and approach work, which would span Jewfish Creek itself. Laborers struggled with eleven-foot-long, lead-heavy crossties of ten-by-twelve-inch oak, chosen in this instance over pine because of its density and superior resistance to the elements. The rails, which were laid at the normal gauge, or width, of four feet eight and one-half inches, also had to be placed by hand, no easy task when each might weigh four hundred pounds.

Still, despite the rigors of the work and the chronic problem of desertion, work on the first leg of construction proceeded steadily. By 1906, the segments from Homestead to Jewfish Creek were nearing completion and work was being readied for Key Largo, when an advance survey party encountered the first of many surprises. As they hacked their way through the

Ida Alice Shourds, whom Flagler married in 1883, two years after Mary's death. At the time of their wedding, Ida Alice was thirty-five, eighteen years Flagler's junior. Photograph courtesy of the Henry Morrison Flagler Museum, Palm Beach, Florida. © Flagler Museum Archives.

(Above) Henry Flagler at twenty-three, in 1853, with his first wife, Mary Harkness (standing), and her sister Isabella Harkness, shortly after Flagler had become a partner in Harkness and Company, a northwestern Ohio grain merchant and distiller. Photograph courtesy of the Henry Morrison Flagler Museum, Palm Beach, Florida. © Flagler Museum Archives.

Satan's Toe, the forty-room mansion overlooking Long Island Sound in Mamaroneck, New York, purchased by Flagler in 1882 for $125,000. Photograph courtesy of the Henry Morrison Flagler Museum, Palm Beach, Florida. © Flagler Museum Archives.

Photograph of Flagler Street in Miami, circa 1904. Photograph courtesy of the Historical Museum of Southern Florida.

The Breakers Hotel in Palm Beach, still standing and looking much the same now as it did then. The original structure was built in 1900 and destroyed by fire in 1903. Its replacement (1906) remains the preeminent hotel in the city. Photograph courtesy of the Henry Morrison Flagler Museum, Palm Beach, Florida. © Flagler Museum Archives.

The magnificent entryway at Whitehall, the seventeen-bedroom Palm Beach home Flagler built for third wife Mary Lily Kenan in 1902. The meticulously restored "marble palace" is now the Flagler Museum. Photograph courtesy of the Henry Morrison Flagler Museum, Palm Beach, Florida. © Flagler Museum Archives.

A postcard of the Long Key Fishing Camp as viewed from the railway station. The legendary camp, once a recreational stop for Florida East Coast executives during the construction of the Key West Railway and later a resort patronized by sports-minded luminaries such as Zane Grey, was obliterated by the 1935 hurricane. Photograph from the collection of Jerry Wilkinson.

Ernest Hemingway in Key West, with his favorite barkeep, "Sloppy Joe" Russell. Photograph courtesy of the Historical Association of Southern Florida, Miami News Collection.

Construction embankments formed by water-borne dredges that dug their way down through the Keys, creating fill for the raised railroad bed as they went. Photograph courtesy of the Henry Morrison Flagler Museum, Palm Beach, Florida. © Flagler Museum Archives.

Workmen engaged in the exhausting task of clearing brush and preparing the right-of-way for the Key West Extension. Photograph courtesy of the Henry Morrison Flagler Museum, Palm Beach, Florida. © Flagler Museum Archives.

Railway cars fitted with wooden freshwater tanks. As there was no source of fresh water in the Keys, thousands of gallons were hauled southward daily from Homestead on the Florida mainland. Photograph courtesy of the Henry Morrison Flagler Museum, Palm Beach, Florida. © Flagler Museum Archives.

One of the many construction camps erected to house workers, materials, and machinery as the Extension progressed down the 103-mile skein of the Florida Keys. Photograph courtesy of the Henry Morrison Flagler Museum, Palm Beach, Florida. © Flagler Museum Archives.

A view of the wooden forms used to pour the concrete arches of the Long Key Viaduct, one of the most graceful railroad bridges ever constructed (and still standing). Photograph courtesy of the Henry Morrison Flagler Museum, Palm Beach, Florida. © Flagler Museum Archives.

The interior of a quartersboat, one of the floating dormitories used to house workmen in the early days of the Extension. After several such craft met disaster during the hurricanes that plagued the project, their use was discontinued. Photograph courtesy of the Henry Morrison Flagler Museum, Palm Beach, Florida. © Flagler Museum Archives.

Construction on the Bahia Honda Bridge, the only steel-fabricated span in the chain of bridges that connected the Keys. The structure was so strong that the pavement of the Overseas Highway was originally laid atop the span. Photograph courtesy of the Henry Morrison Flagler Museum, Palm Beach, Florida. © Flagler Museum Archives.

Concrete support piers for the Pigeon Key portion of the Seven Mile Bridge. Once past Pigeon Key, barely visible in the distance, railroad workers often labored out of sight of land altogether. Photograph courtesy of the Henry Morrison Flagler Museum, Palm Beach, Florida. © Flagler Museum Archives.

(Above right) Chief Engineer Joseph C. Meredith, the "expert in reinforced concrete" who accepted Flagler's original charge to "go to Key West," in July 1904. Actual work on the project began in April 1905. Photograph courtesy of the Henry Morrison Flagler Museum, Palm Beach, Florida. © Flagler Museum Archives.

(Above left) William Krome, who took over as chief construction engineer on the Extension project in 1909 following the unexpected death of Joseph Meredith. Krome had been a member of the original survey party nearly lost in the Florida Everglades during the preliminary stages of the project. Photograph courtesy of the Henry Morrison Flagler Museum, Palm Beach, Florida. © Flagler Museum Archives.

A party on the railroad flats at Key Largo on December 17, 1906. Flagler is in the top hat, with Major General John Brooks to his left and Joseph Parrott to his right. Photograph courtesy of the Henry Morrison Flagler Museum, Palm Beach, Florida. © Flagler Museum Archives.

Mary Lily Kenan, whom Flagler married in 1901 (he was seventy-one, she thirty-four), shortly after the dissolution of his union to Ida Alice, who had been institutionalized in 1897 and declared "insane" by the New York Supreme Court in 1899. Flagler and Mary Lily would remain together until his death in 1913, less than two years after the Key West Extension had been completed. Mary Lily would later marry newspaper-magnate-to-be Robert Bingham, of Louisville, Kentucky. © Flagler Museum Archives.

Henry Morrison Flagler, 1830–1913, Rockefeller partner, railroad builder, Florida pioneer, visionary, in 1909. Photograph courtesy of the Henry Morrison Flagler Museum, Palm Beach, Florida. © Flagler Museum Archives.

last of a thick stand of mangroves just south of the point where the new bridge joined Key Largo, the group stopped short, stunned at what they saw.

A lake stretched out before them, easily a mile across, and nearly as broad as the Key itself. There would be no skirting this previously uncharted body of water, then, and engineer Meredith soon realized that it would be no easy task to cross it, either, for core-drilling revealed that while the lake was scarcely six feet deep, its bottom consisted of a deep bed of peat, deposited undisturbed for eons and far too unstable to support conventional support pilings.

After considerable thought, however, the indefatigable Meredith hit upon a solution. As described by Franklin Wood, a writer for *Moody's Magazine,* a contemporary periodical, Meredith explained to Flagler that they would dredge the clay-like marl, countless tons of it, from the adjoining sea bottom and dump it into the lake to create an embankment across the newly christened Lake Surprise. Though it would be tedious and costly, there seemed little alternative. Finally, Flagler gave his approval to a process that would take almost fifteen months to complete.

The resulting mile-long stretch of track would give rail passengers their first real glimpse of open water, however, and the embankment and the name of the lake remain to this day, giving those who trace Flagler's route by car a similar breathtaking vista and providing inspiration for artists and writers alike:

> The car was up to eighty by the time they rounded the long curve and came up to Jewfish Creek Bridge. The car hurtled up the ramp of the bridge, left the ground briefly, and the undercarriage banged on the other side. The fat man grabbed for the door handle.

Sober as hell now. Adrenaline sober. Night air, going eighty-five through the dark sober.

The young man's foot drove deeper into the accelerator pedal, and he watched the flash of guardrails, saw Lake Surprise appear, the car slewing right, a tire slipping off the edge of the pavement, catching in the shoulder, twisting the wheel from his hands, and he didn't try to recapture it, and the Buick rammed through the guardrail, sailing out into the water. . . .

There was the short flight, the pounding drop, the spray of glass, the sledgehammer to his chest. The warm water of Lake Surprise flooding in. And he lost consciousness.

—James W. Hall, *Under Cover of Daylight*

(W.W. Norton, 1987)

12

Nature's Fury

While the embankment inched its way across Lake Surprise, work down the line continued apace. Camps for workmen were laid out along the Upper Keys, from Key Largo to Long Key (proximate to today's MM 68), a distance of about forty miles, but while supplies could be trundled by rail to the work site at Lake Surprise, everything necessary for the preponderance of the workforce quartered farther along— water, food, medicine, equipment, and building materials— had to be carried around the break in the line by barge or the shallow-draft paddle steamers.

"All there was ready for us was the air to breathe," grumbled one workman, "and that was too thick with mosquitoes to be much good." Added one of the paddleboat captains after his craft had run aground on one of the many shallow reefs abutting Long Key: "[There's] not quite enough water for swimming and too damned much for farming."

Overseeing it all was the quiet but indefatigable Meredith,

the Midwesterner and "concrete expert" whom Flagler had snatched away from one years-long project in Mexico and dropped into the midst of an even greater undertaking in the Florida Keys.

"When I was down there," wrote Edwin Lefevre in *Everybody's*, "Meredith had his headquarters at Knight's Key. In and out of the construction camps he flitted in his launch, his binoculars to his eyes, like a general observing the movements of his troops on the battlefield. You could see telephone poles sticking out of the water in the shallow places, for all the world like lines of skirmishers and scouts."

Meredith's devotion to his job and loyalty to Flagler were as steadfast as any military officer's. For him, as for most of the engineers and supervisors, the successful completion of this singular project meant everything, their sense of duty inextricable from their sense of self. In the same way that NASA engineers half a century later would subsume their individuality into a group effort toward a goal of unquestioned magnitude and significance, the men who designed and oversaw the building of this railroad sensed that as a unit they could accomplish the impossible, and Meredith was the perfect field commander for this team.

He received his orders from the commander in St. Augustine, and conveyed them to his troops with dispatch, zeal, and, when required, creativity. For while the undertaking was guided by Flagler's vision, the commander in chief was wise enough to give his field general free rein when it came to the devising of day-to-day tactics. *Here is the goal,* the wise commander says. *How you achieve it, precisely, is up to you.*

"I was told to make my studies and my estimates," Meredith said. "We had lots of problems to solve, and I was quite a long time at it, and I knew how much [Flagler] desired to see the work rushed, but I never heard . . . one request for haste. When the report was ready, Mr. Parrott and I took it to Mr. Flagler. He heard how we proposed to do it. We stopped before we came to the estimates of cost. And Mr. Flagler stood up and looked at us and said: *'Well, let's get to work!'* . . . Perhaps he felt the occasion called for some comment, for he looked at me and said very quietly: 'I want to see it done before I die.' That is all he said."

By all accounts, Meredith, in his hesitancy to express emotion, even a fit of temper, was much like Flagler himself. So when he closed his interview with *Everybody's* as he did, the sentiments are especially revealing. "Mr. Lefevre, there isn't one of us," he said, "who wouldn't give a year of his life to have Mr. Flagler see the work completed!"

It was a statement that in Meredith's case would prove to be prophetic. For, deep inside, this taciturn man bore a secret that he had divulged to no one, and which would one day have profound consequences for this undertaking.

◻   ◻   ◻

Meanwhile, because the pace of the work for the first half of 1906 had been agonizing, Parrott and Meredith made the decision to carry the project on through the hurricane season, which in the western hemisphere begins to peak in late summer and stretches well into November. It was not a decision to be made lightly in a region where life's milestones were often marked in reference to major storms. Residents of South Florida who might never make reference to such cataclysmic events as world wars still regularly refer to the Storm of '26, or Donna, or Andrew, to give historic context to a birth or death or a family's arrival in these parts.

On the other hand, life in the American tropics has its lotus-eater's aspect. Lulled to a near torpor by the hypnotic press of heat and sun, the repetitive rhythms of pile driver and spike-pounding sledge, and the steady if gentle ocean breezes, men seem to forget what might come churning up out of the summer-cooked waters between Florida and Africa—even today, when, despite the incessant reminders from a modern weather service to stock up and be prepared, long, last-minute lines overwhelm groceries and hardware stores whenever the annual hurricanes finally form and threaten.

And for men not born to the region, which included most of those on the labor gangs in 1906, such a lack of awareness might be understood. Though by nature, not forgiven.

Weather forecasting in those days was nearly nonexistent. Storms on land could be tracked, and advance warning given, reported by telephone and telegraph. But storms approaching the continent by sea were another matter. Communications between the United States and the Caribbean islands are chancy enough in the early twenty-first century. One hundred years ago, they barely existed.

In the Keys, in the late summer, Conchs kept a close eye on the skies to the east and trusted their aches and pains and the odd behavior of livestock and wildlife for tips. (According to novelist and Keys observer Joy Williams, locals know that a hurricane is coming when the colorful red blossoms of the royal poinciana refuse to appear, or when land crabs are seen marching toward higher ground, or when ants climb straight up the walls.) Though the barometer was invented as early as 1643—by an Italian named Torricelli—and though it had been refined considerably by the time Flagler began his march, few modern versions of the instrument had found their way to the labor camps of the Florida Keys in 1906.

Certain enterprising foremen understood the basic principles behind Torricelli's work, however, and while none had gone so far as to erect anything like the thirty-four-foot-tall column of glass he'd originally used to prove his notions, these FEC employees did regularly check their own small tubes full of water in which small wisps of weed rested on the bottom. The principle was simple: if the weed nudged up off the bottom of the tube, it meant that the air pressure was dropping, that bad weather was on the way.

On the evening of October 16, 1906, as it turned out, the weeds in those makeshift barometers all across the work camps on the Keys had begun to rise at an alarming rate. It was an indication, of sorts, for the men stranded on those lonely bits of land, but there was no way they could have reckoned the magnitude of the storm they were in for. It's one of the ironic things about hurricanes: they tend to be fast-moving, tightly packed systems, having nothing much in common with the sprawling fronts of advancing low pressure that usher in periods of miserable weather for most parts of the world, vast systems hundreds of miles broad and deep, lumbering along at five to ten miles per hour, taking their own sweet time in arriving and departing.

If normal storm systems advance like old-fashioned armies, hurricanes are the guerrilla element of weather, zigzagging here and there, now strengthening and speeding in their movement, now appearing to loll, doing everything they can, or so it would seem, to frustrate all those modern instruments arrayed to track them. A densely packed storm such as Andrew, for instance, which blew across the southern tip of the Florida peninsula in 1992, arrived under cover of night as well; a glance at the cerulean eastern skies at sunset would have revealed absolutely nothing out of the ordinary. And when the storm did arrive, early the following morning, it cut a swath that was barely more than twenty miles in width. But what a swath it was.

Residents of Coral Gables, perhaps five miles north of the passing of the storm's eye wall, experienced heavy rain and wind gusts that felled tree limbs, the sort of weather most people might associate with a strong summer thunderstorm. It was the sort of evening to stow the lawn chairs in the garage and have a drink or two inside, feeling cozy while the rain splattered the windows.

But just southward, where the wall of the storm itself passed, matters were inconceivably different. It was almost as if a series of tornadoes had slammed against the shore and marched inland abreast, hundreds of them advancing side by side from just north of Jewfish Creek to just south of Miami proper, the storm roaring like 747 engines doubled, then redoubled again, shearing off roofs and leveling homes, flattening groves and pinelands and utility pylons, flipping vehicles and tossing massive rooftop cooling systems like toys. It is impossible to stand upright in such winds, and even if it were, remaining outside for long would be suicidal.

Kevin Brown or Roger Clemens might manage to throw a fastball in the high nineties, and some major leaguers have suffered fractured skulls when they've been too slow to duck such a pitch. Andrew's winds were running somewhere between 150 and 175 miles an hour. A baseball weighs five ounces. Now try to imagine taking a hurricane-tossed, five-pound clay roof tile in the face. Or think of a jagged hunk of ripped-in-half tin sheeting, ten pounds of it, let's say, Frisbeeing along at 150 miles per hour—try to conceive of what that might do, meeting the human body.

Most of us are spared such prospects, thankfully. Many of us have a hard time even conceiving of them, so far outside the range of normal experience are they. But that is another thing about hurricanes. For those who have never lain prone beneath the passage of such a monster, there is no way of knowing beforehand. And so it was for the men encamped on the northern Keys on October 17, 1906.

Six days before, the hurricane had been just a bothersome tropical storm, lolling off the Windward Islands. By October 15 it had escalated into a hurricane that was blasting Havana. By the next day the storm had taken a sudden turn northward, and by that evening, word had reached men working on the Extension that dire things were in store.

When the men went to bed, the skies were clear, the winds calm, almost unnaturally so, making it seem more humid than usual. And the seabirds seemed fewer in number, too. Because the foremen had their eyes on those rising strips of grass and weed, the boat decks were cleared of loose material, extra lines were lashed to the moored houseboats, stakes and guys on the tents were checked. When everything had been done that it seemed could be done, the men who were scheduled to work rejoined their shifts, driving piles, digging, toting, pounding, and pouring—and those who were spared their labor until morning went to sleep, for a while at least.

William H. Sanders, chief engineer for one of the tugboats in Flagler's fleet, was one of the latter, housed with more than 150 other men on Quartersboat No. 4, one of the floating shanty-style dormitories that was tethered in Hawk's Channel, just offshore from Long Key at MM 67, where work on the line was progressing from south to north. In a *Miami Herald* story, Sanders wrote that by midnight, winds had become so strong on Long Key that all work had to be shut down and the men sent to shelter on Number 4.

Sanders awoke at 6:00 A.M. to find the winds screaming

outside the flimsy walls, and the clumsy houseboat wallowing in huge waves, already taking on water. When he struggled out to the launches the men used to ferry themselves to and from their floating camp, he found every one of the boats useless, their motors soaked by the storm-driven rain. With rowing through the maelstrom out of the question, Sanders retreated inside with the others, hoping that the storm would blow over. As bad as this blow was, it surely couldn't last that long, or so he thought.

At 7:30 A.M., with the winds having risen beyond what any of them had ever experienced, there was a sudden lurch, followed by a sensation of movement that sent a wave of dread through Sanders. The cables that anchored the houseboat had snapped, finally, and the houseboat, designed to be towed along protected waters at near-idle speed, was now being driven southward into the raging Gulf Stream by winds of more than one hundred miles an hour.

As waves smashed over the decking, planks began to give way, one by one. The boat was wallowing now, twisting with every wave crash. Men fought to lash mattresses over the gaping holes in the sides of the houseboat's superstructure, but the winds were too strong.

Just before 9:00 A.M., a mighty gust swept over the craft, peeling the roof away in an instant. The men had a brief, surreal glimpse of nothing but leaden sky where boards once had been, and then, in the next instant, the walls of their shelter were toppled, leaving them exposed to the storm on what amounted to a barren barge-top.

Those who hadn't been swept over the side with the crashing walls fought wildly to find some purchase on the disintegrating decks, but it was hopeless. Sanders saw some of his comrades, some of whom could not swim, others of whom feared the sharks that cruised the deep waters offshore, gulping overdoses of laudanum from the first-aid kits, then lying down to die.

Finally the boat was driven onto a reef, its bottom torn open, its decks pounded to smithereens by the relentless waves. Supervisor Bert Parlin rushed below to warn the men who had taken refuge in the shallow hold. Before he could get a word out, a huge support fell and crushed him. The others in the hold were tangled in the wreckage and drowned.

Some of the men who were still on deck when the boat ran aground were thrown into the sea. The lucky ones, like Sanders, were able to lash themselves to floating planks or timbers.

"One piece of the side wall held together, and about ten of us hung on to it for dear life," Sanders said.

A few feet away, he saw one man clamber onto a huge piece of decking, a two-by-twelve timber nearly forty feet long. As Sanders watched, another plank came screaming through the air toward the man, who looked up in time to take its blow full force. The man's chest split as though it had been cut with a giant pair of shears, Sanders said. In the next moment, a wave swamped his own makeshift raft. When he came up spitting seawater, the man and the planks were gone.

In the aftermath of the storm's passage, in waters that were still raging, the Austro-Hungarian steamer *Jenny* spotted a lone survivor clinging to a chunk of debris and managed to haul him on board. When the man tried to explain what had happened, he found that none of his saviors could speak English. The *Jenny* was about to continue on its course when someone found a fireman deep in the hold who understood enough English to understand that there were hundreds of others lost in the storm-tossed seas.

The fireman relayed the news to the *Jenny*'s captain, who ordered the ship into a full-fledged search. By one-thirty the following morning, the *Jenny* had pulled forty-nine men from the water and delivered them to safety in Key West.

A British steamer, the *Alton*, picked up twenty-four more survivors clinging to a makeshift raft and brought them to port in Savannah. Other survivors were picked up in widely scattered spots; one man was delivered by a freighter docking in Mobile; another was carried to port in Liverpool. More turned up in Galveston, New York, London, and Buenos Aires.

Other workmen and other boats suffered less fortunate fates. Some survivors told the story of a cement barge tender named Mullin, who refused to leave his craft when it was blown away from its anchorage. There was a generator on board, and Mullin valiantly kept his boiler stoked and his lights ablaze for hours "like a Coney Island steamboat," until the waves finally pulled the barge under and Mullin was drowned.

Another man, swept out to sea on one of the barges used to ferry water to various points along the right-of-way, managed to grab a wrench and loosen the bolts that held one of the big cypress water tanks to the deck of his disintegrating craft. He jumped into the tank, pulled its lid tight above him, and rode out the storm like a bug inside a hollow cork. Several days later the tank washed up near Nassau, its occupant badly dehydrated, but alive.

A pair of mechanics named Kelly and Kennedy were clinging to the deck of another floundering quartersboat when an unmanned barge floated by. According to one of the other men on the houseboat, Kelly shouted to his friend, "I like the looks of that barge, don't you?"

Kennedy must have liked its looks as well, for the two of them jumped from the houseboat to the deck of the barge. Moments later the pair disappeared in the boiling seas and were presumed lost. A week later they were discovered far out in the shipping channels, nearly dead from lack of food and water. The two recovered, though, and returned to work on the Extension.

There were many bizarre tales told following the disaster. One pair of men tossed into the sea from Quartersboat No. 4 was a father-son team. As they floundered about in the water, they caught sight of a trunk floating nearby and swam desperately to it. They held tight to the makeshift life preserver for a time, but eventually the pounding of the waves and the wind exhausted them. First the seas washed the father away, and soon after, the son felt the trunk torn away from his hands as well.

The son was fortunate enough to grab hold of a plank, though, and managed to keep himself afloat until he was rescued the following day. When he was finally delivered to safety, the son told a railroad official the heart-wrenching story of the loss of his father.

"Tell me your name again, son," said his listener.

The son did so.

The official smiled then, and clapped the young man on the shoulder. "You can relax. Your father's safe. He told the same story when he was brought in a couple of hours ago."

The tale was a heartening one, but in all more than 125 men died, and because record-keeping of the day was so slipshod, the toll was rumored to be as high as two hundred. When Joseph Meredith arrived to survey the devastation, he was understandably shaken, but his public statements reflected the resolve that had won him the position of project manager to begin with.

"No man has any business connected with this work who can't stand grief," he said, vowing to go forward.

Flagler's response to Meredith's insistence was just as terse: "Go ahead," was the word sent down from St. Augustine.

13

Duly Noted

Despite Meredith's public proclamation, the death toll produced by the storm was staggering. Although a certain amount of risk accompanies any heavy construction project, even to this day, the loss of a fellow worker is a major blow to any labor team, especially one such as that working the Overseas Railroad, where camaraderie was underscored by the relative isolation, long hours, and clear sense of purpose reflected by their employers. Imagine, then, the loss of not one, or two, or several compatriots, but 125 fellow workers—nearly 5 percent of all those hard at work when the storm swept ashore—all of them vanished overnight, and many of the bodies disappeared forever beneath the featureless seas.

If any man had harbored doubts about the wisdom of sticking with the job until now, he needed no further persuasion. With most of the nearly three-thousand-man workforce idled for almost twelve months while the storm damage was surveyed, and the basic supply and staging infrastructure reassembled and repaired, the rate of defections skyrocketed.

To make matters worse, those fleeing back northward carried with them grim tales that newspapers and eager labor prosecutors were quick to embroider. "Some of the boys came down here evidently to get their passage paid without any intention of working for the company," one official complained to a reporter for the *Brooklyn Eagle*. "A number got employment in Miami with local concerns and we are out the cost of their passage. It is these fellows that often stop men at the railroad station and on their way to the works and fill their ears with all sorts of false stories."

But when affidavits were filed by some former workers alleging that railway security officials locked the dock gates at the port of Miami, preventing their return to the U.S. mainland, public outrage reached a crescendo: in March of 1907, a federal grand jury in New York issued its initial indictment charging the Florida East Coast Railway, including Meredith and Krome, along with its New York labor agents, under "slave labor" statutes.

Stung by the charges and frustrated by the setbacks to his project, Flagler mounted a counterattack. Photographs of the Miami docks and rail terminal facilities were circulated in an attempt to show that the claims of workers being "denied re-entry to United States soil" were patently ludicrous. Flagler's attorneys also managed to solicit an affidavit from a Catholic priest who had toured the Keys during the pre-hurricane construction phase, stating that he had seen no signs of forced-labor operations.

Flagler would also issue a detailed statement defending the railroad's labor-management practices, one intended to reassure the world at large that the FEC was indeed a benevolent employer. While the document might have intended to present the company's practices in a favorable light, given the attitudes of the time, reading it today offers a telling insight into Flagler's pragmatic nature:

> The white laborers were recruited in New York, Philadelphia, Pittsburgh and other northern cities, and while most of it was of a class not used to the rough kind of work needed in railroad construction, and included a large portion of undesirable types, it was thought they could be educated up to our requirements. . . . the Negroes who remained or who came in later, were of a much better grade than the first recruits, and for the clearing of the heavy jungle along the right of way the Negro was far superior to the whites.
>
> The most satisfactory results were secured through the letting of contract work, the contracts being let to individuals who would hire ten or twenty men to help them, the price being fixed in advance for the work to be done. The contractors seldom failed to make good wages and a profit on the work. The station men . . . however, particularly the Negroes, seldom kept their earning very long owing to their tendency to gamble and spend their money in liquor. . . . During a part of 1907, Spanish laborers from Cuba, attracted by the high scale of wages paid on the concrete construction at Long Key Viaduct, began to enlist and since then have formed a considerable element of the labor force,

> *and these men have been a generally satisfactory*
> *type of laborer. . . . They are "stayers" and hoard*
> *their earnings carefully, seldom leaving the work for*
> *"layoffs" or junketing trips.*

Such comments are fairly compelling evidence that whatever racial attitudes Flagler may have held in his heart of hearts, he was principally concerned with one issue: getting this particular job done. In an earlier letter to Joseph Parrott, Flagler had expressed his concerns that his men be well treated on the job. He had read a *New York Herald* interview with a member of the Panama Canal Commission outlining certain aspects of care of a labor force under tropical conditions, and wanted to send Parrott a heads-up: "While we have no condition such as exists on the Isthmus, it occurs to me that whoever is in charge of that work should look carefully into the question of everything pertaining to the health and proper care of the force we employ."

And in an aside during his defense of the FEC's hiring practices, he grumbled about another of his pet bugaboos: "There were strict rules prohibiting the introduction of liquors into the camps, but it was undoubtedly brought in by the men themselves. There was this noticeable difference, however: the Negroes were regular, but moderate drinkers while the whites who drank would get on a spree of several days before returning to work."

Despite Flagler's attempts to elicit a more sympathetic response from the public, newspapers outside the realm of his influence were skeptical. When a flare-up of a nagging liver condition put Flagler back in the hospital during the summer of 1907, the *New York Journal* declared him on his deathbed, the dream of an Overseas Railroad slipping away as well:

It was the "railroad that goes to sea," and it has been Mr. Flagler's ambition for years to leave that to posterity as a monument to himself. He had personally supervised all of its workings; but when success seemed most sure, when hundreds of miles of the great railroad that was to connect Miami with Key West, and make the latter virtually a land city, the project had to be abandoned—temporarily, say his friends; probably for all time, say those engineers. . . . Since early spring, not a wheel has turned in the railroad's construction, not a boatload of concrete or iron been taken South, and the thousands of men who were employed in the work have sadly packed their kit and drifted back to New York, leaving the half-built giant railroad practically deserted. . . . Broken in spirit and badly weakened from overwork and worry incident to the strain of raising funds, Mr. Flagler took to his bed in St. Augustine last April. He did not mend rapidly and his physician advised his removal to a cooler climate. Early in the summer he was taken to his summer home in Mamaroneck . . . but was later removed to Bretton Woods, N.H., when symptoms of a general nervous breakdown asserted themselves.

However, reports of Flagler's demise were greatly exaggerated, as the saying goes. Even as the *Journal* account was being published, crews were back at work, rebuilding work camps and replacing roadbed and rails that had been washed out by the storm.

By October of 1907, 2,500 men were back on the job, and

work camps had been established as far south as Knight's Key, at MM 41, more than halfway to the final destination. As the process continued, one major change was implemented: as a result of the tragedy involving Quartersboat No. 4, there would be no more floating dormitories. All the camp buildings would be constructed to meet hurricane-resistant standards, and all would have secure foundations on dry land.

A reporter for the *Chicago Daily News* made a tour of the post-hurricane camps and offered a perspective that heartened Flagler and his managers: "I doubt if there could be found better conditions for the common laborer in any engineering work that exists along this extension," wrote F. S. Spofford. "I spent two days in these camps, mingled with the men, ate their food, inspected their quarters, and I must say that on every hand were evidences of the greatest consideration for their welfare." After a detailed rendering of the typical camp menu, Spofford concluded his dispatch with a reprise of the often-cited FEC assertion that the percentage of illness among workers had been much lower than that experienced by members of the U.S. Army.

In response to charges that men had been prevented from leaving the camps, or had been prevented from passing company gates on the pier into Miami until they had paid back their transportation debts, William Krome issued a statement that insisted, "All Resident Engineers . . . were clearly instructed that no difference how a man's account stood in regards to his indebtedness to the Company, that if he wished to leave, no effort could be made to detain him, the only detention being that he was required to pay $1.50 return fare to Miami."

Press reports of the day continued to bat back and forth the issue of Flagler as a modern-day slavemaster, however, and federal prosecutors were as resolute in moving their case against the FEC forward through the New York district courts as Flagler was in extending his railroad southward. When the case, based on an 1866 slavery law, finally went to trial in November of 1908, things went badly for the prosecution, its efforts undermined largely by the presentation of a series of distinctly unreliable witnesses. Inside of a week, Flagler's lawyers petitioned for, and were awarded, a directed verdict of acquittal by the trial judge.

Flagler, though pleased, reiterated his suspicions that the entire matter had been part of a government conspiracy, engendered by various frustrations over the years in making antitrust charges stick against Standard Oil. (As early as 1878, lawsuits seeking to break the company's stranglehold on the oil business were being filed, and Flagler, who had arranged most of the company's contracts, found increasing amounts of his time devoted to such rearguard actions, duty which he found loathsome. Though a U.S. Supreme Court decision in 1911 would finally mandate the dissolution of the intricate network of the Standard Oil Trust, the effects were negligible. Subsidiary stocks remained in the hands of Rockefeller, Flagler, and others who had controlled the company, and less than a year later it was estimated that the $663-million value of the former Standard Oil of New Jersey had actually skyrocketed an additional $222 million.)

Flagler attributed much of the government's animosity to a personal vendetta being carried out against him by President Theodore Roosevelt, whom he had once supported in a bid for the governorship of New York. Once elected to that post, however, Roosevelt had moved quickly to pass legislation taxing corporate franchises, an act that Flagler deemed a personal betrayal. In the years that followed, Flagler's disdain for Roosevelt grew to gargantuan proportions, as his private correspondence bears out. "I have no command of the English language that enables me to express my feelings regarding Mr. Roosevelt," Flagler says in one letter unearthed by biographer David Leon Chandler. "He is shit."

Despite the lingering venom, the court victory had given Flagler's plans a significant boost. On November 25, 1908, just five days after the government had its case quashed, Flagler wrote to Elihu Root, "I have today decided to resume work on the Key West Extension. We will have trains running to Key West on January 1, 1910."

14

On Toward Key West

All the while that Flagler was laid low and the case against the FEC was wending its way through the federal court system, work had nonetheless gone forward down the Keys. Some sixteen miles of track already laid on Key Largo had been washed out by the hurricane of 1906 and had to be rebuilt. After that, Tavernier Creek, the channel between Key Largo and Plantation Key at MM 90, was spanned by a low-lying trestle about half a mile in length.

Plantation Key, referred to as "Long Island" by the original road builders, constituted another five-mile stretch of "web-footed" work for Meredith and his team: clearing the gnarly mangrove thickets and palmetto scrub; blasting in the shallow waters offshore for limestone fill to build up the roadbed; tamping, grading, and leveling; laying ties and rails; and filling the gaps with coarse riprap, followed every step of the way by the growling supply trains and the relentless clouds of mosquitoes that drank the blood of the workers as resolutely as the workers guzzled fresh water carted down in the huge cypress tanks.

After Plantation Key, and the building of another low-lying trestle to span Snake Creek, engineers had to contend with two tiny dots of land originally known as the Umbrella Keys. In his article describing the possibility of this route, Jefferson Browne had written, "Between some islands, short trestles will be necessary, but some of the passages could be filled with loose rock," and a decade later, Meredith found himself doing that very thing.

Rather than bridge the narrow channel separating the Umbrellas, Meredith adopted Browne's elegant if environmentally dubious suggestion. He simply increased the volume of quarrying and fill dredging, and thereby joined the two islands into one whole, which would eventually become known as Windley Key (MM 86). Because the Umbrellas' formation offered a relatively thick layer of fossilized limestone and coral upthrust from the surrounding sea, and because there was no one around to protest, the tiny keys were quarried extensively for fill, both during the construction of the railroad and later, during work on the adjoining highway.

Any land in South Florida that rises high enough to escape regular dousing by salt water is likely to become a haven to hardwood outcroppings, and the Umbrellas were no exception. Work crews encountered dense stands of mahogany, torchwood, and ironwood (so hard that carpenters despaired of driving nails into its lumber) standing in their path, with their deep-burrowing roots and steely trunks providing yet another unforeseen challenge. The rare trees were felled and the dense wood used for fuel, but this was no logging operation. Whatever of the hardwood hammock wasn't in the way survived for the most part, though quarrying the intricately patterned rock on Windley Key would never stop altogether.

Just south across another narrow channel from the newly formed Windley Key lay what the engineers and the surveyors referred to as Matecumbe, designating it as one fourteen-mile-long island, though a conservative estimate was that six of those miles consisted of marsh, creeks, and an open channel that separated what soon became known as Upper and Lower Matecumbe Keys. Though the work on the Matecumbes was difficult, it was essentially work of a familiar sort, no different from the chop, blast, and fill operations that had been in use all the way from Homestead south, and progress across the Matecumbes and its neighbor to the south, Long Key, was rapid.

While supplies to the southern camps, Long Key among them, were ferried in by barge and steamer, the FEC was using a pair of well-traveled 4-4-0 steam locomotives built in the 1880s to deliver supplies, equipment, and material, not to mention a fresh supply of labor, as the line moved steadily southward from Miami. A considerable amount of specialized equipment was built on site by FEC employees pulled from their normal postings: rolling camp cars with sleeping and dining facilities, machine-shop cars, portable blacksmith cars, and an oversized handcar dubbed "Bull Moose" by the workers.

When the barge-mounted excavators that Meredith had devised for working the shallows ran out of water to float upon, he had them lifted by crane aboard flatcars and hauled to the next workable spot. Because the roadbed at times traversed right-of-way that was some distance inland, a fair amount of temporary spur track was laid, often on wooden trestles that tacked out over the shallow reefs to where the limestone marl was being dredged from the sea bottom.

At one point Flagler's general manager, Joseph Parrott, who

served as staging master to Meredith and his corps of construction engineers, chartered every American steamship available for hire on the East Coast to haul hundreds of thousands of tons of coal, steel, lumber, machine tools, food, and medical supplies. The crushed rock needed to finish the roadbed filled some eighty steamer loads by itself.

The focal point of this stage of the work lay just south of the Matecumbes, on Long Key, where the first of the truly awe-inspiring bridges would have to be constructed, spanning an uninterrupted 2.68 miles of water before reaching a speck of land now known as Conch Key. Whereas all the bridges in the hundred miles or so of construction up to this point consisted principally of low-lying trestlework, this was the first undertaking that would tax the ingenuity of Meredith and his associates. Preparations for work on the Long Key Viaduct, as it was called, had been under way as early as 1906, but when the hurricane swept in, most of the preliminary work was washed away, and engineers had to begin again on what would be their most demanding task to date.

Long Key, the jumping-off point for the viaduct, was in many ways the perfect place to locate a staging area. At the south end of the island (MM 65.5) the choking mangroves recede, supplanted by natural sandy beaches, and the surrounding shallows extend seaward for a half a mile or more, offering striking views of an endless dappled turquoise and blue waterscape. Bathers, flats fishermen, and the just plain curious could walk through waters no more than knee-deep until they had scarcely any sight of land.

And it was here, nearly two-thirds of the way to his destination, that Flagler built a series of quaint, screened cottages for his white-collar workers, in an area that the writer Joy Williams would characterize as a virtual paradise: "Everywhere there is water, water that becomes sky, the shadows of rays like clouds moving across the blue. Water loves light. The light changes. Dawn and sunset break. Thunderclouds mass. The water is black, emerald, azure, sheer, and the vault of sky becomes the vault of water. Flocks of egrets fly bone-white across that impossible interstice . . . lovely litanies of colors and creatures, fishes and birds."

Priscilla Coe Pyfrom, whose father, Clarence Coe, served as one of Flagler's chief bridge engineers, described some of the professionals who came to live in the area camps and work with her father on the project: "Civil engineers started out at $125 per month . . . with room and board and, if needed, free medical attention. At that time they were glad to get a job. . . . Many of these fresh-faced young engineers had college debts to pay off." And many were more than happy to avail themselves of home-cooked meals and hospitality from Coe's wife, who assumed the role of mother-in-absentia, often cooking dinners and hosting weekend gatherings to bolster the spirits of those hardworking young men so far from home.

"It was still, with all its inconveniences, a wonderful life," Priscilla Coe recalled. "We were given almost perfect conditions to grow up in: a mother at home to take care of our daily needs, a father who, though not rich, always had a job he adored doing with an income above average. . . . [W]e spent winters in Florida and summers on an Iowa farm. Who could ask for anything more?"

15

The Signature Bridge

O ne of those who asked for "more," of course, was Henry Flagler. By this point in time, he had narrowed the focus of his project upon the completion of the first of the great bridges that would have to cross a truly formidable body of water on the way to Key West.

In Flagler's eyes, the Long Key Viaduct was the essential link that would culminate the first phase of his "impossible" project. Once the viaduct was finished, it would connect with sixteen miles of track that had already been laid on Grassy Key, the next in the southwestward chain below Long Key, where yet another in a series of unforeseen developments had forced Flagler to announce a temporary terminus for the project.

One of Flagler's key assumptions to the financial success of the Key West Extension had always been the establishment of a deep-water port in the southernmost city, one which could accommodate the vast amount of oceangoing steamship traffic on its way to and from the proposed Panama Canal. But Flagler's plans were dealt a formidable blow in 1908 when the U.S. Navy refused to grant permission to dredge the waters of Key West Harbor to build up the huge dock area.

Though Flagler suspected that the influence of his old nemesis Roosevelt was responsible for the setback, his response was typically pragmatic. He dispatched a team of engineers to study the waters south of Knight's Key, off Key Vaca, ostensibly to see whether or not his deep-water port might be relocated, possibly replacing Key West as the railway's ultimate destination. It was news that both stunned and delighted Middle Keys residents, whose dreams of supplanting Key West as the most influential city in all of Florida were suddenly given life.

Meanwhile, Meredith, Coe, and others were hard at work on forging this last link between Knight's Key and the nearly one hundred miles of track that snaked its way down the archipelago to languish just to the north. Renderings for the new bridge, with its more than 180 steel-reinforced concrete arches rising thirty-five feet above the water, resembled nothing so much as a great Roman aqueduct marching across the sea. The bridge was so striking in appearance that Flagler would come to call it his favorite of all those built on the line.

Spokesmen for the company were fond of quoting the statistics derived from the building of the viaduct: 286,000 barrels of cement were required for the arches and the pilings, much of it a special underwater-hardening type that would have to be imported from Germany. There were 177,000 cubic yards of crushed rock hauled down, some of it to be used to fill the inner chambers of the arches and create the actual railroad bed, the rest to be blended with the cement and 106,000 cubic yards of sand (enough to cover all of Miami's famed South Beach) to make concrete. There would be 612,000 feet of pilings sunk to create the underpinnings of the span, over 5,000 tons of steel reinforcing rod laced through the super-structure of the arches, and more than 2,500,000 feet of timbers used to build the massive forms.

Yet the numbers, great as they are, give little sense of the difficulty that was involved in building that great chain of archways. "It is perfectly simple," Flagler's brave words surely echoed in the minds of the builders as they struggled. "All you have to do is build one concrete arch, and then another. . . ."

These arches were being built, however, not on dry land, but in the middle of an ocean that varied in depth anywhere from ten to thirty feet. For every one of those more than 180 arches, each spanning a distance of over fifty feet, a flotilla of barges and work boats had to be moved into place, whereupon the intricate process would begin.

William Mayo Venable, FEC division engineer, was the first to detail the nature of the work in a 1907 article for *Engineering Record,* a process that was later amplified upon by Dan Gallagher in a 1995 pamphlet, "Pigeon Key and the Seven Mile Bridge," describing similar work farther down the line.

First, pilings had to be driven through sand and bottom muck into the bedrock to serve eventually as anchors for the arches themselves and to prevent side-to-side movement of the supports that might be occasioned by tidal surges. Sometimes a piling would reach bedrock quickly. Other times it would mean hours of deafening, precarious work, exacerbated by storms and squalls; currents that were always shifting the barges, tilting the heavy, steam-driven equipment out of place; and all the other surprises that nature seemed fond of springing in a place where such work had never before been undertaken.

At times, a piling being driven just a few feet away from one that had been stabilized would sink and sink into a sea bottom that suddenly seemed as porous as quicksand. Exasperated engineers could either keep on pounding, hoping to strike rock before reaching China, or simply give up, move their rigs a bit to one side or the other, and hope that this time they wouldn't hit another hidden pocket.

After the pilings were finally secured, a rectangular wooden form or cofferdam was constructed about the cluster of pilings, the top of the form poking a few feet above the surface of the water, its base secured in the sea bottom and shored up by submerged sandbags piled around its perimeter. Once this preliminary form was secured in place, a layer of the special underwater cement (used because Venable and others feared that American cement might not resist the action of seawater; the reports of various cement tests take up several shelves in the Flagler museum holdings) was poured into its bottom, surrounding the pilings and forming a watertight seal two to five feet thick, depending on the size of the pier, and resting directly upon the ocean floor.

After the seal had been allowed to set for two or three days, water and debris would then be sucked out of the form by large, barge-mounted pumps, and leaks in the form patched, so that workers could build a second, more refined form inside the first. This form within a form would receive the latticelike network of reinforcement steel woven around and about the pilings and then the whole would be filled with concrete, to form a pedestal of sorts, its top projecting ten feet or so out of the water, and studded with a veritable hydra-head of steel reinforcement rod ends left waving in the air, to eventually tie into the last piece of the process.

In the final step, and after a week or so of hardening, a pair of these pedestals would be joined by yet another form, one

The American frontier closes at last: the first scheduled train traverses the Key West Railway Extension from Miami, arriving in Key West on January 22, 1912. Photograph courtesy of the Henry Morrison Flagler Museum, Palm Beach, Florida. © Flagler Museum Archives.

Celebration! An aging Flagler, who vowed to ride his own rails to Key West before he died, is helped past a schoolchildren's choir, part of the festivities arranged by local officials to celebrate the achievement. Photograph courtesy of the Henry Morrison Flagler Museum, Palm Beach, Florida. © Flagler Museum Archives.

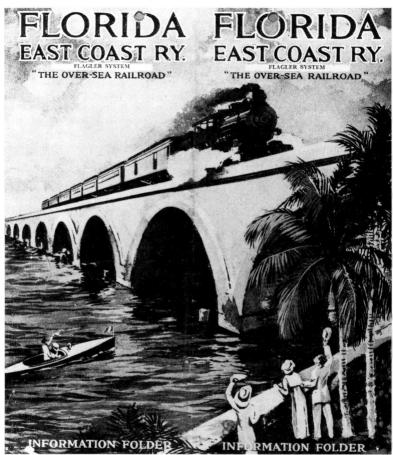

Much promotional literature engendered by the FEC in the wake of the Extension's completion featured fanciful renderings of a locomotive steaming across the famed Long Key Viaduct. Ironically enough, the view was accessible only by boat.
Photograph courtesy of the Henry Morrison Flagler Museum, Palm Beach, Florida.
© Flagler Museum Archives.

This stylized rendering served as the FEC logo until the devastating Labor Day hurricane of 1935. The company, already in financial straits, would go into receivership in the storm's aftermath. The Extension right-of-way was sold to the state for a pittance; it would later serve as the route for the Overseas Highway. Photograph courtesy of the Henry Morrison Flagler Museum, Palm Beach, Florida. © Flagler Museum Archives.

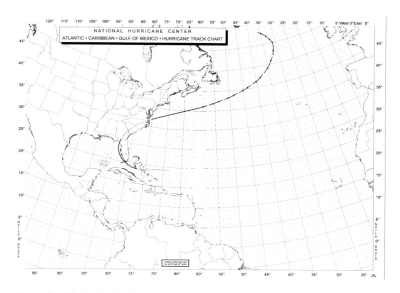

The track of the Labor Day hurricane of September 3, 1935, believed to be the most powerful storm ever to strike U.S. shores. Winds reached more than 200 mph as the storm passed over the middle Florida Keys. Chart courtesy of NOAA.

The remains of an FEC rescue train, dispatched from Miami late in the afternoon of September 2. Only "Old 447," the train's engine, managed to stay upright as the great storm-driven tidal surge swept across Matecumbe Key, carrying hundreds to their deaths. Photograph courtesy of the Historical Museum of Southern Florida.

Rescue workers, including Ernest Hemingway, described the scene at the storm's Ground Zero in hellish terms, the landscape stripped and blasted clean for more than forty miles. Photograph courtesy of the Henry Morrison Flagler Museum, Palm Beach, Florida. © Flagler Museum Archives.

This view of the devastation looks northward at Long Key, where Flagler once pondered ending his Overseas Railroad. Photograph courtesy of the Henry Morrison Flagler Museum, Palm Beach, Florida. © Flagler Museum Archive.

An aerial shot of the damage at Snake Creek, where the rescue train was delayed for more than an hour when a wind-blown cable became snagged in the locomotive super-structure. Photograph from the collection of Jerry Wilkinson.

Corpses of hurricane victims awaiting Coast Guard cutter 212 for transport to Miami. Photograph from the collection of Jerry Wilkinson.

The body of a victim is loaded aboard a rescue boat at Snake Creek. Rescuers were shocked at the speed with which bodies decomposed in the late summer heat and humidity. Photograph courtesy of the Historical Association of Southern Florida, Miami News Collection.

A military guard performs a gun salute beside coffins stacked for a funeral pyre erected at Snake Creek. While officials had first vowed to bury all the bodies of the veterans who died in the storm, first at Arlington, then in Miami, health officials decreed that the rapidly decomposing remains would have to be burned to prevent the spread of disease. Photograph from the collection of Jerry Wilkinson.

A vivid testament to the unprecedented fury of the 1935 storm is seen in these two pictures of the Millionaire's Club on Matecumbe Key, which was built by eleven stock exchange millionaires. Above, the club after the lesser storm of 1933, and below, the club completely devastated in the wake of the 1935 storm. Photographs from the collection of Jerry Wilkinson.

built on land and transported to the site by barge, and this in the shape of an arch. Once the arch form had been set in place, more concrete would be poured atop it. The result was one gracefully curved, fifty-five-foot link in what would eventually become a breathtaking chain of arches across the sea.

However, because there was always the possibility that the contraction of the drying concrete span might place some stress on the piers at either end, the builders would skip over the next link in line, "hopscotching" their way along, joining one pair of piers with an arch, then leaving a gap, then joining the next pair of piers, and so on. From a distance it might have looked as if a series of giant croquet wickets had been plunked down in the ocean at regular intervals.

Once the alternating series of links had completely cured, the builders would go back and fill in the remaining gaps in the chain, thereby ensuring that the structure could not twist or weaken itself as it dried, and avoiding any interwoven stresses that could weaken the entire chain.

After all the arch-bottoms had been poured, wooden side-forms would be added along the chain, and the final pouring of concrete could be undertaken. That last step created the sides and top of the arch itself, and formed the actual base upon which ties and track would be laid. Once the final pour of concrete had hardened, the sides and the arch-bottoms, or "arch-rings," as they were called, could be removed and used again.

It was tedious work that often went on around the clock, with generators and steam engines grinding incessantly, and involving an entire corps of divers outfitted in primitive Jules Verne equipment, fumbling about in murky and turbulent waters shot through with tricky currents that, despite Jefferson Browne's observation—"the water in which the trestling would be built would be no rougher than that of any of our larger rivers"—could pull a man and his hundred pounds of weights out to sea in the blink of an eye.

As many as eight hundred men were working on the Long Key Viaduct at one time, according to Venable, most of them housed in the Long Key Camp, which was beginning to resemble a city in its own right. There was even a bakery on the premises, Venable noted, churning out one thousand loaves a day to feed the hungry workers.

Despite the daunting nature of the work, by January of 1908, less than a year and a half after construction had resumed in the hurricane's aftermath, the Long Key Viaduct was completed. Flagler was delighted with the bridge's mirage-like appearance, but he had also realized that there was no way to appreciate its beauty and singular aspect as a passenger on the train itself.

As a result, Flagler had a photographer set up equipment on a barge out to sea, then had one of the first trains to cross the span halted midway across so that a series of dramatic photographs could be taken—proof positive, in his estimation, that the impossible had become reality. "This viaduct alone," went the caption to one company rendering of the day, "is a monument to a whole lifetime of constructive skill and enterprise."

It wasn't long before a likeness of a smoke-belching train atop an aqueduct that spanned an ocean would become the enduring trademark of the entire Florida East Coast Railway system. To this day, the dramatic icon pops up throughout the Keys, festooning local businesses and working its way into the work of folk artists and painters inspired by the romance of the Keys.

"The track is thirty-one feet above high water," enthused a travel writer who accompanied one of the first groups to make the crossing, "so that passengers in the railway trains may sit in the windows of Pullman cars in serenity and have an opportunity of seeing how the Atlantic Ocean looks in a gale." Later, in a letter to his friend Hemingway, John Dos Passos would call the trip a "dream journey."

Shortly after the viaduct was completed, Flagler had his palm-studded work camp on Long Key converted to a fishing camp, and built a terminal at which guests could depart the main line and ride a half-mile, narrow-gauge spur through a tunnel beneath the tracks and out to those inviting accommodations on the sandy beaches. He also had a proper hotel constructed directly on the water, and the compound would eventually become a world-renowned destination for the more sports-minded of the affluent set drawn to the Flagler resorts.

In time, Zane Grey would take up a recurring part-time residency at the Long Key Fishing Camp, working on his popular Westerns on a schedule he described as "an hour of writing in the morning, followed up by ten of fishing." Grey's enthusiasm and versatility as a sportsman remains legendary in the Keys. He is generally given credit for popularizing the practice of angling for sailfish on light tackle, and was one of the first championship fishermen to advocate the practice of releasing his trophy catches.

The completion of the Long Key Viaduct marked a significant milestone for Flagler. It meant that passengers could board a train at the Miami station and travel all the way to Knight's Key (MM 47), for 106 of the 153 miles between that point and the proposed destination of Key West. Or they might board one of the FEC's Pullman cars in Jacksonville and ride in comfort down 477 miles of Florida coastline. Even though much of the hardest work remained unfinished on the railroad across the sea, enough had been accomplished that it seemed

Flagler's "one arch after another" theory was not as out-landish as it originally seemed.

In February of 1908, regular passenger service between Miami and Knight's Key began, though before he could insti-tute it Flagler had to contend with a demon from his past. The timetable issued by the railroad for the new service enumer-ated stops at stations at Long Key, Grassy Key, Key Vaca, and Long Key Dock. But for years the train made an additional stop on Key Vaca that was never listed in the timetable.

A group of Conch settlers whose holdings blocked the extension of Flagler's right-of-way down the narrow island had banded together, though in this case their object was not to drive up the price of the land that Flagler hoped to cross; they simply wanted their own stop along the new line. After protracted debate, Flagler gave in, perhaps moved by the fact that for once it wasn't greed that was getting in the way of his railroad's march. In any case, a deal was struck, and once a week, or whenever the Conchs hung out a signal flag, the trains of the FEC would grind to a halt at the unlisted station of Vaca.

It might have been an annoyance for the great man, but Flagler was always one to keep his sights on the larger picture. After considering his engineer's surveys of the nearby reefs, he'd gone ahead with the dredging of a nineteen-foot-deep channel dug to allow better oceangoing access to Key Vaca and Knight's Key, and instituted regular steamship service by the Peninsular and Occidental to meet the trains. Passengers could debark and go on to Key West by steamer, or take advantage of direct steamer passage to Havana, just 115 miles to the south.

Before long, through service was established between New York City and Knight's Key, Florida. Mondays through Saturdays, frigid Northern passengers could board the *New York and Florida Special* at 2:10 P.M. of a murky and snowbound Manhattan afternoon. At 7:30 A.M. on the third day following, they could wake up in a berth of a Pullman car and raise a shade to look out the window at a stretch of blue ocean framed by glittering skies and waving palms, a steamship awaiting at a nearby dock. Six hours later the most adventuresome of those passengers could find themselves steaming beneath the lowering aspect of Morro Castle into breathtaking Havana harbor.

As had been the case wherever Flagler's line had reached a terminus, however temporary, a veritable building frenzy began. Two work camps had been erected during the rush to finish Long Key Viaduct, along with engineering offices, a camp hospital, machine shops, a locomotive repair facility, executives' cottages, recreational facilities, and a power plant.

The burgeoning steamship service required docks, a ferry terminal, a customs office, a post office, and a floating hotel to accommodate passengers during the inevitable layovers occasioned by schedule delays and heavy weather. One worker housed in Camp No. 10 on Key Vaca was overheard to say, "Building this railroad has become a regular marathon." The remark struck a chord in his fellow workers, who dubbed their camp "Marathon," the name by which the nearby town, the second-largest in the Keys, is known today.

These were heady times in the Middle Keys, to be sure, but it was also a time of great uncertainty for Lower Keys and Key West residents, as well as for those who wondered if Flagler would finally give up on his impossible task and be content to stick with the docks near Marathon as his terminus.

In a *New York Times* interview, Joseph Parrott made emphatic denials that the company was abandoning its plans to reach Key West. In that interview he also forcefully denied wild rumors that project manager Joseph Meredith had been fired and that a notorious swindler—former Army captain Oberlin M. Carter, who had embezzled millions from the Savannah Harbor Works project—had been hired to take his place.

Parrott's assurances that the company was simply gathering steam for a final push did not convince everyone, however. WILL FLAGLER QUIT? went the headline of one local editorial, and residents south and west of Knight's Key debated long and hard whether Flagler, his flag firmly planted in the Keys now, his steamers sailing proudly on to Cuba, would be fool enough to tackle the next insurmountable obstacle in his way: seven miles of open water that separated the railway's present terminus and the next thing resembling dry land in the chain.

16

Seven Miles of Hell

As usual, those who doubted Henry Flagler's resolve were bound to be disappointed. His legal troubles had been a significant distraction (his legal fees alone were estimated at more than eighty thousand dollars), his health had not been good, and even Meredith's key assistant, William Krome, had begged off the project for a time, citing his own exhaustion. But the downtime following the completion of the Long Key Viaduct allowed Flagler an opportunity not only to rest and deal with the bothersome ancillary matters, but also to stabilize passenger operations on the completed leg of the Extension, and to undertake another massive staging operation for what would be the most exhausting portion of the entire project.

Following the court's dismissal of the Department of Justice charges in November of 1908, Flagler ordered supplies to be stockpiled on the Lower Keys and his crews back to work at full speed, with their sights set on the most mighty task to date: the Seven Mile Bridge.

When word reached Key West that work was under way again, the response was giddy. "Key West looks northward to the fast approaching bands of steel which will bind her to the mainland and dreams of the not far distant day when she shall be a large and bustling and important city, the metropolis of the Southern Seas," gushed an eager correspondent for the hometown *Key West Citizen*.

And if Flagler was not quite so fulsome, he had plenty to be pleased with in early 1909. Work on the Long Key Fishing Camp had been completed, and guests were already clamoring to register before its scheduled opening in early January. Work on the first bridge pier south of Knight's Key was completed in February, and his vice president and confidant, William H. Beardsley, informed him in early April that actual costs of construction for the first quarter were running less than half of what project supervisor Meredith had estimated.

But yet again, and just as matters were finally looking up, tragedy stuck. The slightly built Meredith, as it turns out, was diabetic, a condition he had revealed to no one. With his frail constitution taxed from the outset by the rigor of the project and the demands of the climate, he had finally begun to weaken early in 1909, the end result of four years of grueling work beneath the tropical sun.

While Meredith had not complained, Flagler had noted that his project manager was not operating with his normal vigor, and sensed that something was wrong. For weeks he had been urging Meredith to go north for a rest, and had even coaxed William Krome back to work on the project to fill in while Meredith took some time off. Though others on the job site understood that Meredith had been feeling rocky from time to time, no one was prepared for what was to happen.

On the fourteenth of April, Flagler had paid Meredith a visit in Marathon, where they discussed the progress of the work. In his diary, Flagler noted that Meredith was looking better than he had in a while, and supposed that his supervisor was on the mend. When Flagler left to return to St. Augustine, Meredith went to Key West on business, but hardly had he arrived than he was taken ill. When fellow workers realized just how weak and disoriented he was, a doctor was summoned, who ordered Meredith taken immediately back to Marathon, where he could be transferred to the company's hospital in Miami by train.

By the time he arrived in Marathon, however, Meredith was too weak to be moved, and it was decided that he be taken on to Miami by boat. It was Tuesday morning before the ship carrying Meredith arrived in Miami, and by then he had slipped into what doctors discovered was a diabetic coma. He died early that same afternoon, on April 20, 1909.

It was a blow that struck at the morale of the entire workforce and echoed all the way to Flagler's offices in St. Augustine. For more than four years, ever since Flagler had uttered the fateful order "Go to Key West," Meredith had been the human engine who drove the otherworldly effort.

In a ceremony worthy of a war hero, Meredith was laid to rest in a Miami cemetery, and Flagler would keep a news clipping concerning his project supervisor's death in his effects until the day he himself died. During the services, all Florida East Coast Railway operations, including rail traffic, were halted for five minutes. His gravestone was decorated with a bronze plaque with wording approved by Flagler:

In memory of Joseph Carroll Meredith, Chief Engineer in the construction of the Key West Extension of the Florida East Coast Railway, who died at his post of duty, April 20, 1909. This memorial is erected by the railway company in appreciation of his skill, fidelity, and devotion in this last and greatest work of his life.

At that moment, with his chief engineer gone, Flagler could easily have been forgiven for accepting the advice of the naysayers. What lay ahead, said one writer, was more than daunting: "So far as can be seen, it is a matter of launching a railroad straight at the blank horizon of the Atlantic." Perhaps Flagler himself wavered: should he cut his losses, retreat to Knight's Key, and base his operations there?

But retreat—or surrender—was not in Henry Flagler's nature. He turned to Meredith's second-in-command, William J. Krome, the strapping young civil engineer who had led the ill-fated mapping expeditions through the wilds near Cape Sable; without hesitation, Flagler offered him the project manager's job. Krome, who had returned to work only two weeks before, accepted on the spot.

Though Krome was only thirty-two, he had been well seasoned, working alongside Meredith every step of the way. "Mr. Krome is a very efficient man," Flagler wrote to a friend on April 22, "and we have no anxieties about his being able to prosecute the work successfully."

Even if Flagler was as confident as he sounded, however, common sense would have dictated that Krome, a graduate of the University of Illinois and Cornell and a member of the respected Society of American Engineers, would have his own private reservations. The "work" most immediately at hand involved the building of a bridge more than 35,000 feet long, or nearly four times the length of the daunting Long Key

Viaduct. With the necessary approach work figured in, the span was almost nine miles long from tip to tip. Nothing remotely like it had ever been attempted before, and Krome would have been a fool not to consider the myriad possibilities for disaster that lay in his path.

One of the first difficult decisions that Krome had to make was whether or not to continue work on the bridge and the remaining track down the Lower Keys through the upcoming hurricane season. Prudence suggested he send his crews home once summer came and take up the work again late in the year, once the neighboring seas had cooled and the engine fueling storm development had essentially been shut down.

But with Flagler now seventy-nine, and rumors of his precarious health still circulating, Krome worried that his chief might not live to see the project finished. Whether it was simply devotion to the cause and to his employer, or whether Krome might also have doubted that the project would continue with Flagler out of the picture, is difficult to say. In any case, Krome finally made his announcement: despite the risks, the project would continue through the summer and fall.

In a letter to Flagler, Krome let his apprehensions be known, but also laid out his contingency plans, in case the worst should come to be:

> No man has ever passed through one of the West Indian hurricanes and boasted that he had no fear of it. Indeed, lack of fear is dangerous. The responsibility resting upon the engineers for the safety of the men and for the preservation of equipment is heavy. There is no harbor along the entire line of the grade that is safe from hurricane. We must be ready for it when it comes; we must have the workmen

*well in hand to prevent panic. We must have done
all we could to save our machinery and camp outfit.
We have found it more economical to sink our floating equipment in the most protected waters and
raise it and repair it when the storm has passed.*

Flagler took Krome's warnings to heart. Telegraph wires
were strung from Miami "down the grade" to the work camps
so that timely weather bulletins could be relayed to supervisors. Men were no longer housed—even temporarily—in the
precarious floating dormitories, but in sturdily constructed
barracks on shore. Women and dependents were required to
leave the keys sites by August. An evacuation plan was established, with rescue trains ready to be mobilized at a moment's
notice throughout hurricane season.

And meanwhile the arduous work continued, exacerbated
by the draining heat of summer and the unrelenting swarms
of mosquitoes. While preliminary groundwork on the approaches to the Seven Mile Bridge had gotten under way as
early as 1906, the building of the span itself had been post-
poned while engineers studied how best to go about it.

With Meredith gone, the task fell to Krome and his cadre of
engineers, chief among them C. S. Coe, who had recently been
promoted to Division Engineer. Coe, a Minnesota native and
graduate of the University of Minnesota, had worked as a railroad construction engineer specializing in bridges for eighteen
years when Flagler tapped him to join the Extension project.
In his time with the FEC, Coe had performed the yeoman's
work demanded of everyone on the project, including oversight on the building of bridges over Jewfish Creek, Tavernier
Creek, Snake Creek, and the broad channel connecting Lower

Matecumbe Key and Long Key. But it was his work on the Long Key Viaduct that focused full attention on his abilities.

Coe had been assigned responsibility for the survey of the waters in the Knight's Key Harbor, where the temporary terminus and steamship docking facilities were constructed, and he was in charge of building the extension and four thousand feet of complicated trestlework that curved out from the main line so that rail passengers could be delivered directly to the dockside. He had also been put in charge of the design and construction of one of Flagler's pet projects, the Long Key Fishing Camp. When the time came to tackle the mammoth Seven Mile Bridge project, Coe was Krome's choice to oversee the task.

After considerable study and survey work, Coe and Krome determined that varying water depths and other factors dictated that the massive bridge would actually be built in four rather distinct sections, and in two completely different styles. The first section, the Knight's Key Trestle, would link the tip of Knight's Key with Pigeon Key, a dot of land about a mile offshore, where an auxiliary work camp could be established and staging undertaken for the next link in the crossing. A short second section, referred to as the Pigeon Key Trestle, would be built over the shallows west of Pigeon Key to join with the fourteen-thousand-foot main segment, the Moser Channel Bridge, which would include a 253-foot section that could be swung aside to allow the passage of tall ship traffic between the Atlantic Ocean and the waters of Florida Bay. The final two-mile section of the span, designated as the Pacet Channel Viaduct, would reconnect with land on Little Duck Key, at MM 40.

This multistaged design meant that the bridge would first

curve a bit to the north instead of marching straight across the void, thereby missing some of the deepest water separating Knight's Key and Little Duck. Interestingly, it was a strategy that had been suggested by Jefferson Browne in his 1894 essay.

The first three segments were to be constructed of steel deck girders laid atop a series of 546 concrete support piers, while the Pacet Channel segment, crossing waters that were barely deep enough for barges to float, was to be built in the spandrel arch style—210 of them—similar to those of the Long Key Viaduct.

The first step in the actual construction was for surveyors to complete the painstaking task of determining the locations of the piers and arch supports, another job complicated by the fact that all this work was taking place at sea. Since surveyors' transits could not be used to sight levels on bobbing boats, platform after platform had to be built at five-hundred-foot intervals across the seven miles of ocean, each of them firmly anchored in bedrock. The platforms themselves had to be located out of the way, about two hundred feet south and east of the line the actual construction would take.

When the surveyors had completed their calculations and triangulated the necessary positions, they used flag and hand signals to direct workers in the precise placement of the initial piling for each of the 746 bridge supports. Once the initial pilings for each support had been set and their alignment checked (a tall wooden tower was constructed on Pigeon Key for this purpose), the other pilings could be plotted and driven down to bedrock.

All of the heavy construction equipment used at the time was powered by steam engines, which, in comparison to still-to-be-developed diesel, electrical, and hydraulic tools, were massive, noisy, and notoriously undependable. The image of a vintage steam locomotive thundering down a mountain pass, belching clouds of black smoke in its wake, might be a romantic vision in the modern mind, compared to the passage of a diesel or electrically powered engine, but imagine having to work on a yawing, heaving, pile-driving barge from dawn till dusk, never more than a few feet away from one of those roaring, foul-smelling behemoths.

Such engines also consumed great quantities of coal and fresh water, all of which had to be transported over great distances to the work site, along with most of the rest of the materials used in the construction. Each one of the bridge piers required enough sand, gravel, cement, lumber, and steel to fill a single five-masted schooner.

Most of the cement used for construction above the tide line was of domestic origin, much of it from New York. The cement used for submarine applications was brought from Alsace. Steel girders and track segments were fabricated in Pittsburgh, pine lumber came from Georgia and Florida, hardwoods from sources in the Midwest. Even the rock and sand used to mix concrete was imported from elsewhere because of concerns that the salt content of the local deposits would cause corrosion of the reinforcement rods.

And yet the very size and scope of such a task produces its own inescapable inertia: As any child knows, once the lumbering engine is set in motion, "I think I can" soon becomes "I know I can," and Flagler's crews were soon working at a record pace.

Dredges scooped out the muck and marl and sandy sea bottom in areas too shallow for barges to maneuver, and dumped their spoils to create man-made islands still visible from the bridge today. Next, the pilings—anywhere from seventeen to twenty-one of them, depending on the size of the pier—were driven into the exposed limestone cap formed by the remains of vast, ancient coral reefs.

Crews found the top layer of the limestone to be a relatively stable crust that varied from three to six feet in depth. Below that, however, the deposits were softer and pocked here and there by caves and layers of sand. In some places, pile drivers were unable to penetrate the cap rock more than a foot. In others, the operators would have to give up searching for a solid footing and hope that the crust they'd driven would hold. Some of the pilings might be twenty feet from top to bottom, others might be fifty.

Once the pilings were in place, helmeted divers were called in to aid in setting the outer forms, or cofferdams, then pouring the bottom cement seals, which were left to harden for four days. Once the seals had set, the water and debris were pumped out of the cement-bottomed cofferdams, and wooden molds were lowered inside and secured to the base. While the molds were different for the support piers than for the spandrel arch bases, the principle involved was the same. Workmen descended inside these inner molds to assemble the reinforcement rods, the ends of which would project above the tops of the forms to tie into the next section.

Then derricks mounted on nearby cement-mixing barges would transfer concrete to be poured into the forms, bucket by agonizing bucket. Once this step was completed, the rest of the work would take place above the waterline.

In the case of the spandrel arches forming the south end of the Seven Mile Bridge, the arch-ring molds could now be set in place, and the rest of the more complicated process would proceed. For the steel girder supports, what remained was a simpler matter. A last prefabricated form was set atop the pier that poked out of the water, and after more reinforcing rods were assembled inside, the form, sometimes rising as high as twenty feet above the water, was filled with concrete. After the concrete was allowed to harden for a few weeks, the forms were removed to be used again, and the steel support spans could be laid from pier to pier.

The support spans used in the Moser Channel were eighty feet long and nine feet high, so that the final roadbed lay nearly thirty feet above the ocean. Each of the huge girders weighed nineteen tons and had to be ferried to the work site on barges pulled by tugs, then lifted into place by massive Derrick Boom No. 9, a seventy-foot monster mounted on yet other barges.

When a pair of the spans had been secured to iron footings atop the piers, they were joined together by a series of cross braces and angle braces riveted into place. Once the spans were rigidly braced, crossties could be bolted atop them, and the rails fixed into place.

An account of an inspection tour over the sixty- to eighty-foot elevated spans by one of the workers gives a sense of the breathtaking nature of the work: "After dinner walked the girders (fourteen to eighteen inches wide). Ties were laid for a bit, but farther out there is nothing to walk on but the girders about twenty-six feet above the waters," he wrote, before adding this understated afterthought: "It makes for some bad walking when the wind is blowing."

If no problems intervened, crews could normally complete as many as four support piers in a week, and ironworkers could join as many spans in the same time frame. During one shining week in 1909, with support piers ready and everything falling perfectly into place, crews set a record by joining twelve spans, a rate that surely must have pleased "the Chief," as Flagler had become known.

In an entry in his own diary dated June 29, 1909, Flagler

noted that the last of the concrete support piers for the
Knight's Key Viaduct had been completed, and by August, he
wrote to an associate that good progress had been made on the
Bahia Honda Bridge as well. Rails would soon reach his ultimate destination, Flagler predicted:

"Unless we are hindered by unfavorable weather or delays
in the delivery of steel for bridges, I think we shall have the
track ready to run trains to Key West on or before the first of
February [1910]."

Meanwhile, Flagler was hardly content to sit idly while
work progressed. He kept a close eye on every decision, firing
off dozens of memoranda and letters to Krome and his supervisors, often wondering if some task might be undertaken in a
more efficient or economical manner.

In August of 1909, for instance, Flagler wrote to Krome,
referring to the tides that were passing beneath those bridges
they were struggling to build: "The thought often occurs to me
that there is an immense amount of power going to waste
between the piers on the viaduct and the three bridges, and I
am often wondering whether it would not be feasible to utilize
this power by making electricity with floodgates. Can't you rig
something up?"

Krome's reply was a masterful rendition of dodging and
double-talk, managing to reassure his employer that the man-
ufacture of electrical power by harnessing the tides was a matter well worth looking into but one that would have to wait a
bit while other fish were fried. Whether Flagler was fully pal-
liated or not is unclear, but Krome was permitted to turn his
attentions back to the task of reaching Key West in time to
meet his boss's brave predictions.

Given the way things had been proceeding, most observers

felt that for once Flagler was not exaggerating when he said the line would reach Key West by the first of the year. But, once again, nature had its own plans.

On October 10, with work on the Knight's Key Viaduct very nearly wrapped up, and Flagler ready to think about having timetables to Key West printed, the U.S. Weather Bureau reported that a hurricane had struck western Cuba and was now approaching the Florida Keys. The Weather Bureau had established an observation station on Sand Key, about eight miles southwest of Key West, where steadily dropping barometric readings suggested that the storm was strengthening far more rapidly than had been anticipated.

By 8:30 A.M. on the morning of October 11, winds had reached seventy-five miles per hour on Sand Key, and weather observers were ordered to take shelter in a nearby lighthouse. Shortly thereafter, the anemometer cups were sheared away by winds estimated at one hundred miles per hour. Trees and fences were flattened, and the tides had engulfed the entire island. At ten-thirty, a massive surge carried the weather station away into the seas, where it was shattered and sunk.

W. R. Hawkins, resident civil engineer at Marathon from 1909 to 1912, kept meticulous notes on his experiences with the FEC. On the evening of October 10, he read the bulletin issued by the Weather Service in Washington, and, after spreading the word, copied the message down in his careful hand:

"NE storm warning. Tropical disturbance west of Havana, moving NW by N with apparently increasing intensity. Increasing NE wind tonight along gulf coast of central and southern Florida."

There was no mention of the Keys in the bulletin, but

Hawkins understood the implications. There was only one way to get from Havana to the mainland coast, and he and the others were in the middle of that path.

Early the next morning, the worst had happened. "Raining and blowing hard at 5 A.M.," Hawkins wrote. "Tides came very high. Nearly everyone on Boot Key [where a maintenance facility was located] went to Marathon [on nearby Key Vaca]. Had to hold on to the rails a good part of the way walking up the track to freight yard."

And it is little wonder the men had to make their way along like crabs, bent over and clinging to the tracks. Hawkins had no access to a wind gauge, but the Weather Bureau estimated that winds had reached 125 miles per hour by the time they hit Marathon, where Hawkins and some three thousand others were quartered. On Knight's Key, gusts were powerful enough to lift five of the nineteen-ton girder spans that had not yet been fully secured to their support piers, tossing them into the sea.

Earle Hartridge, one of bridge-builder C. S. Coe's assistant engineers, described his ordeal in a letter: "We experienced one of the worst hurricanes at our Pigeon Key Camp, the center passing over us—we thought we would never survive it but we did by making a kind of crater in the top of the cement stored in one of the warehouses, where we rode it out, although the warehouse itself was blown away."

Not everyone was so lucky. J. H. Brown was aboard House-boat Beta, and, according to Hawkins, refused to go ashore when he and his companions were ordered to high ground at Marathon. "The houseboat was blown off," Hawkins writes, "and poor old Brown was killed or drowned."

Thirteen more men, the crew of the tugboat *Sybill*, died

when their craft was driven down, but the shelters held fast for the men who had heeded the orders to come on land. There were no confirmed reports of other deaths.

Workers who ventured out after the worst had passed expressed amazement at the storm's caprice. One man walked the completed portion of the bridge to where the massive, tons-heavy girders had been ripped away into the sea, and found himself staring in amazement at a keg of nails that still perched undisturbed at the edge of one of the now-empty piers.

While Krome's precautions undoubtedly had saved the lives of many, the damage to the railway itself was massive. Some accounts were telegraphed from Miami to the rest of the nation, describing the extension as a total loss and claiming that forty men had actually died.

Flagler was quick to dispute these claims, estimating that he had suffered no more than $200,000 in damages. Other surveys suggested that the costs would rise into the millions: more than forty miles of roadbed and track had been washed away in the Upper Keys, with boulders weighing as much as ten tons dislodged by the surge of storm waters.

Still, Flagler was not deterred. In a letter to James Ingraham, he admitted that the company's losses had been formidable, "but to the general public we must all keep a stiff upper lip and admit nothing." To other associates, he urged, "My recommendation is to hoist Key West's flag high, keep it waving and let it bear the inscription, 'Nil Desperandum.' "

In an aside to Ingraham, however, Flagler gave some indication that his respect for the power of the storms was growing: "I say to you in confidence that I don't want to invest a dollar in building improvements on any of the Keys."

It seems that Flagler's narrowing of focus and his flag-waving paid off. By November 8, less than a month after the storm had supposedly wiped out the line, the company had somehow managed to restore passenger service to Knight's Key, with construction and supply trains following in the wake.

17

Learning Curve

As it turned out, Krome and Coe made one heartening discovery in the storm's aftermath: all of the piers and viaduct supports that had so recently been set for the Seven Mile Bridge, as well as those of the great Long Key span, had emerged from the battering of winds and tides unscathed.

It was considered a major victory, for while the engineering team had designed those support structures to withstand forces four times as great as any storm had ever produced in the Keys, experiences of the last five years had taught everyone involved with the project that there was a vast difference between figures on paper and what was actually experienced in the lovely but treacherous Keys. It was with a great sense of relief, then, that work resumed on the pilings and arch supports across the broad channel.

What had happened in the Upper Keys, along the stretch of the railway constructed according to more conventional standards, was another matter, however. Local residents and federal consulting engineers alike had from the first maintained that adding fill instead of bridges wherever possible had created the potential for disaster, and the effects of the hurricane of 1909 proved the critics right.

A glance at the map illustrates the issue. Between the southern tip of the Florida peninsula and the island of Cuba, a distance of more than 150 miles, lie the waters of the Florida Straits, which constitute a broad channel connecting the Atlantic Ocean with the Gulf of Mexico. Another 150 miles or so southwest of Cuba lies the Yucatán Peninsula of Mexico.

Geologists theorize that the entire distance separating the United States from Mexico was once an unbroken stretch of land, and the Florida Keys that dot the waters between Cuba and the mainland United States are its vestigial remnants, all that remain after eons of erosion by wind and storm and tide. And in the existence of these vestiges, attractive as they may be to tourists, developers, and railroad builders, lies the problem, especially when a monster storm comes boiling up from the Caribbean, pushing a wall of water ten or twenty or thirty feet high ahead of it.

Until the railroad builders arrived, those storm surges, formidable as they might be, would course along the natural passageways that eon after eon of preceding surges had gradually cut and deepened through the reefs. Even tidal waves tend to follow the path of least resistance, and much of that force would simply be shunted into tremendous rivers of water surging through the natural cuts northwestward toward the Gulf. Once the storm had passed on and the waters were released from the driving force of the winds, the currents reversed direction and poured back through the channels separating the Keys toward the Atlantic. If something were to be suddenly dropped in the path of these surges, obviously, there would be hell to pay.

The problem was serious enough in the Middle to Lower Keys, where the passage to the Gulf was relatively unimpeded, the whole process of tide reversal spread over vast, open seas. But in the Upper Keys, the storm-driven waters would barely pass through the channels before they piled up against the mainland. With nowhere to go, a tidal wave would roar back across the Upper Keys much more quickly and with much greater force than to southward. The picture is a simplified one, but it lays out the nature of the wave action quite clearly.

When the railroad builders chose to close off natural channels, then, they were building what were essentially unreinforced dikes, stretches of earth and gravel perhaps twenty feet wide and four to six feet high, capped by wooden ties and railroad tracks. Such structures would have to stand against the forces that had carved through the Keys since time immemorial.

The result in 1909, of course, was no contest. Forty miles of track were swept away, boulders the size of automobiles carried off like pebbles. Even some of the barges that had been purposely sunk to avoid their being dashed to pieces on the shoals had been slung across the sea bottom and destroyed, or buried beneath tons of tide-shifted sands.

In this way, then, the Keys themselves taught Flagler and his men some valuable lessons. Where six miles of bridges were originally planned, eighteen miles would now be built to allow for natural tidal flow.

In addition, the design of the railroad bed itself would be changed. Originally the practice was to dredge or blast for marine marl in the shallows just offshore, then use that material as the substratum for a thick layer of crushed rock and gravel brought in from elsewhere and piled up until the proper grade height was achieved. In the aftermath of the storm, Krome and his men made two key observations. The first was that the worst effects of the storm had been created not by the advancing waves that slammed ashore on the Keys, but by the tides rushing back out to sea. They also noted that while the imported rock and gravel had been swept away by the tides, along with the ties and rails, quite often that substratum of limestone marl remained intact.

Flagler himself pondered these findings and, on October 16, wrote to FEC operations manager J. P. Beckwith with an unprecedented suggestion. They would reverse the normal roadbed building process from now on, Flagler said, first laying down a base of rock and then covering it with the slimy mixture brought up from the nearby sea bottom. Once in contact with the air, the marl hardened, forming a kind of natural concrete seal that was not unlike the very constitution of the neighboring landmass upon which the roadbed was built, and one far more likely to resist the action of storm-tossed waves than the riprap normally used atop grade.

Flagler's theory proved correct and saved the railroad untold thousands over the ensuing years, but it was only the beginning of an ongoing process of construction research required by the singular project. Engineers were also concerned about the effects of the salt spray upon the steel deck spans and reinforcing girders. An FEC paint laboratory established in Marathon would determine that no paint could be devised that would provide protection against the local conditions for more than two years. As a result, painting crews would be required to work year-round on a job that would never cease, so long as the railroad stood.

The issue of wave and salt action on the concrete support

piers was also investigated, but there the news was more optimistic. The results gained by observing concrete samples submerged in laboratory tanks, and by the detailed record-keeping on piers already put in place, suggested that while the concrete might erode slightly after being set (up to one-sixteenth of an inch), the process would eventually stop, impeded by the formation of algae and other marine-aided deposits that effectively sealed the concrete.

By the beginning of 1910, then, work had resumed full force, all of it guided by a directive Flagler issued to Krome in the aftermath of the hurricane: "From this time on let us bear in mind that what we need is permanent construction, even though it costs a little more money and takes considerably more time. I would rather be two years completing the line to Key West and have it permanent, than to have a repetition of the disaster of the hurricane of the 11th."

Permanent construction of the sort Flagler was referring to would not come cheaply, however. Early in 1910, Flagler wrote to John Carrere, designer of the Ponce de Leon Hotel in St. Augustine, that repairing the damage caused by the hurricane had actually cost him $1 million, and reiterated that it had taught him a valuable lesson about upgrading the quality of the work. He estimated that it would require at least another $9 million to push the track to Key West, a figure that did not include the costs of a terminal and docking facilities.

"As it is," he told Carrere, "I have got to finish the work out of my income and I cannot expect to live long enough to do that."

As a result, Flagler confided, he had taken an unprecedented step, a secret "known to but one other person." After talking the matter over with Krome, Flagler had gone into debt for the first time since the project had begun, authorizing J. P. Morgan to issue $10 million in bonds to underwrite the completion of the route. It was a momentous step for Flagler, a clear admission that the project had become more formidable than even he had reckoned.

"I should feel less anxiety," he concluded, "were it not for the Governmental raid upon the Standard Oil Company [from which Flagler derived his most dependable flow of income]. Perhaps this will clear up in a few months and enable me to take a little more hopeful view of the future."

⌧   ⌧   ⌧

While marine crews labored to complete the various sections of the Seven Mile Bridge, land-based crews pushed track southward along the Lower Keys—Little Duck, Ohio, Missouri—and Key West–based crews worked northward over Stock Island, Boca Chica, Big Coppit, Sugarloaf, Cudjoe, Summerland, Ramrod, Little Torch, and Big Pine Keys, some forty miles of work altogether.

Weekly work reports filed by Coe give some sense of the nature of the effort:

. . . for the week ending January 29, 1910.

Pile Driver #1 drove dolphins [preliminary pilings] *in Moser Channel up to Thursday when she moved over to the rock pile and replaced the dolphins torn out by the blow of the 21st.* [Note that it was not only hurricane weather that plagued the project.] *On Friday she moved on line of Pigeon Key Trestle #2 and on Saturday drove 13 piles in the trestle.*

Pile driver #11 worked in Moser Channel all

*week replacing the anchor and fenders piles of the
piers up to Saturday when she commenced driving
on Pier #54.*

*Pump Barge #12 cleaned off the old seals ready to
allow the cofferdams to be replaced.*

*Catamaran #3 set five cofferdams for the week.
The last cofferdam set being #45. The cofferdams
have fitted over the old seals very nicely so far.* [The
references are to repair of pier bases not yet completed when the hurricane hit.]

*. . . Derrick Barge #9 completed setting the steel
on Knight's Key Bridge. Settling the last span on
Wednesday. She set the first span in Moser Channel
on Friday and set two more on Saturday, making 4
for the week.*

*50 recruits were received from Key West on
Thursday which brings our force up to the limit.*

The last is a telling comment, for it suggests that regardless
of all calamity, and in spite of the rigors of the work, Flagler
was by now having relatively little difficulty in finding men
willing to join his effort. Another progress report issued at
about the same time breaks down the workforce headquar-
tered on Pigeon Key in greater detail:

Engineering and Accounting	*18*
Foremen	*13*
Subsistence [Cooks & other service]	*28*
Launchmen	*12*
Skilled labor	*27*
Common labor	*189*
Total	287

The 1910 U.S. Census provided a further breakdown of the workforce stationed at Pigeon Key: 61 of the men came from 28 different states, with New York providing the most, at 12. The other 150 men tallied came from a welter of countries, including 77 from Spain, 33 from Grand Cayman, and 13 from Ireland. Only 5 black workers were listed.

The breakdown offers some insight into the FEC's evolving hiring practices. The Spanish workers had always been "stayers" in Flagler's eyes, and as word got back to their countrymen about improved living and working conditions on the project, more and more of them came to sign on. Native Caymaners, likewise accustomed to the climate and the insects, had also come to constitute a significant part of the workforce.

In addition, the camp on Pigeon Key offered accommodations that were far superior to those of the early outposts built on the Upper Keys. Most men were quartered in sturdy bunkhouses and had access to an infirmary on site. Each of the dormitories housed up to sixty-four men, and each had a reading and recreation room with electric lights powered by a generator. There was a domestic staff that washed and changed linens once a week.

A writer visiting the camps for the *Railroad Gazette* noted that while mosquitoes were "large and fierce," all the bunkhouses and porches were screened. "The men get up at 5:00 in the morning, take a bath . . . have breakfast at 5:30, work from 6:00 to 11:00, have dinner, go back to work at 12:00 and knock off at 5:00, with supper from 5:30 to 6. Sundays are rest days." Many workers reported that conditions at the camp were superior to those at their own homes.

Also, supervisors had learned to look the other way on some matters of policy. The derelict freighter *Senator* was now anchored permanently in Boot Key Harbor, just offshore from the busy work camps on Pigeon and Long Keys, and was doing a thriving business as a bar and house of prostitution. And whether or not Flagler cared much for the idea, by now the FEC was making no efforts to send the *Senator* packing.

Wherever they were from, and whatever their duties and their off-duty proclivities, the workforce by this point seemed to have rallied about the drive to complete the project so that the Chief could "ride his own iron" to Key West while he was still able. Though the Seven Mile Bridge remained to be finished, and a last great channel was yet to be crossed, the race to finish was on.

18

Railroad Builder
Overboard

For bridge-building supervisor C. S. Coe, the race
nearly ended at this point. Because so many key projects were
now under way simultaneously, Coe had been assigned his
own motor launch, staffed at all times by a captain and a
marine engineer.

According to Coe's daughter, Priscilla, the craft had two
compartments, one where the captain and engineer spent their
time, and another aft, where Coe kept a kind of floating office
he used to review plans and the like, and where he might steal
a nap as they traveled between job sites, not an unimportant
benefit for a man who was now working very nearly around
the clock.

One morning, as the launch was traveling south from
Pigeon Key toward the work on the Bahia Honda Bridge, the
last of the great challenges before him, Coe glanced up from
his paperwork, realizing that he had time to make an interme-

diary stop at a troublesome project site along the way. Because there was no means of communication between the two compartments, he was forced to leave his own cabin and make his way forward along a narrow passage on the boat's deck.

He was halfway to the front when the launch piled into a wave and sent Coe sprawling against the handrail. As he clutched frantically, the handrail gave way and he toppled headlong into the water.

He came up choking on brine and calling after the boat, but the captain and the engineer had their eyes ahead, engaged in a boisterous conversation of their own, and Coe's cries were lost in the roar of the launch's noisy motor. Coe was left to shout and wave his arms to no avail, watching helplessly as his launch disappeared. In moments he found himself alone in a desolate sea, miles from the sight of land.

For Coe, who could barely swim, the thought of reaching shore was out of the question. But he had learned to tread water, and to float. Forcing himself not to panic, he kicked off his shoes and worked himself onto his back, doing his best to keep his head above the swells.

As he worked and willed himself to stay afloat, Coe kept his mind on a story that had always stayed with him, in large part because of his very fear of just what had befallen him. A friend had once told Coe about being on board a small ferryboat that was crossing Lake Michigan. The ferry had sunk and the friend had found himself flailing about in the frigid waters, along with his wife and his young daughter, neither of whom could swim.

Coe's friend had reasoned firmly but reassuringly with his wife and daughter, telling them not to panic and to simply keep their hands on his shoulders. He insisted that he could keep the three of them afloat, and if they would simply trust him, and stay calm, they would be saved.

Coe's friend had been successful, as it turned out, and the message behind the story burned fiercely in Coe's brain as he floated there in the lonely waters. He had certain things going for him, he told himself. There was plenty of daylight left, and compared to the choppy and frigid Lake Michigan, he might as well have been bobbing in a calm, if enormous, tub of bathwater.

Still, as the sound of the launch's motor receded, he could feel the panic beginning to surge up inside him. Coe, who had seen more than one man die in his years on this endeavor, knew just how desolate his surroundings were, and how quickly these crystal waters could become deadly.

The irony of it all, though. The stupidity. If he'd simply been content to wait until they'd reached their intended destination, he could easily have scheduled his stop on the way back . . .

. . . and then he stopped, realizing that something had changed. Instead of receding, the sounds of the distant launch motor seemed to be growing louder. Coe listened intently, not certain whether he could trust his ears. Sound played strange tricks as it crossed these waters, another lesson he had learned.

But soon enough he was sure. The boat had turned around, the distant rumble now a roar once again. Coe was back to treading water, doing his best to thrust himself up where he could get a glimpse of the launch, or give his men sight of him.

And then, as a swell lifted him up and he saw the approaching craft, his engineer waving and shouting, relief washed over him. In moments he felt the hand of his engineer on his own, and then he was being pulled up to safety.

As he would learn, what his men had been debating, as Coe

bobbed in the waters and the boat left him behind, was whether or not the boss might want to put in at a certain stop along the way and have a look at a bit of work that had been giving them all fits.

When the engineer had made his way back to Coe's cabin to check, he'd been amazed to find no one there. That was when he had noticed the sheared-off handrail and realized what must have happened.

⊞   ⊞   ⊞

Coe's survival was miraculous, but what he would live to accomplish was equally amazing, at least to most observers. For the remainder of 1910, work continued relentlessly on the remaining miles of track down the Lower Keys as well as upon the completion of the Seven Mile Bridge. However, the difficulties that plagued the project did not abate.

Jefferson Browne, the original booster of this route, had been accurate when he wrote that construction of the bridges spanning the Seven Mile and Bahia Honda channels would be the most difficult and expensive portion of the work. But, he maintained, the final section, from Bahia Honda to Key West, "presents no difficult problems . . . the surface of the islands is as level and smooth as a ballroom floor." Easy for Browne to say. But as Flagler's engineers had learned, nothing in the Keys was simple.

While Coe wrestled with the difficulties of bridging the Bahia Honda, others encountered a problem they had hardly expected in the tropics: forest fire. Little Duck Key, where the southern end of the Bahia Honda Bridge was to make landfall, marks the beginning of what are usually referred to as the Lower Keys. While the Upper Keys are formed of coral outcroppings and support relatively little natural vegetation, those south of the Bahia Honda Channel are low-lying formations of solid limestone that can be classed as true islands, and where varieties of scrub brush, cactus, palm, and even stands of southern pine are found.

Big Pine Key (MM 32), eight miles by two, is the second largest in the entire chain after Key Largo, and, at the time of the railroad's approach, contained pockets of fresh water just under its rocky surface. Moreover, Big Pine not only featured a sizable pine forest (which Browne had theorized could be leveled to furnish all the wood necessary for the railroad ties), but one that housed the extant population of the singular Keys deer.

At the time the railroad workers arrived, local Conchs had already perfected an early form of the "controlled burn," though this version had nothing to do with fire prevention. The Conchs were simply trying to vary the source of protein in their diet by setting small blazes that would flush the timid, two-and-a-half-foot-tall deer out of hiding and into snare traps or across the sights of a rifle.

In addition, other settlers had established charcoal-making kilns on the island, where the flourishing buttonwood trees could be converted into a readily transported source of fuel. Unpredictable wind gusts and the general carelessness of the kiln operators often resulted in unintended conflagrations, and if that meant trouble for a bunch of railroad-building inter-lopers, there weren't many Conchs who were going to lose sleep about it. Though the possibility of fire had never been an issue during all the years of work on the Key West Extension, now Krome and his crews were often forced to stop work or literally flee for their lives when a wind-whipped wildfire raged across the right-of-way.

Nor was every settler happy at the encroachment of builders

who "mobilized their digging machines along the waterfront, streaked the islands with railroad grades from end to end and boosted the prices of real estate by leaps and bounds." There was Montenegrin Nicholas Mackovtich, for instance, a recluse who set booby-trapped spring guns all about his property and refused to speak with anyone who tried to reason with him or buy him out. And, in an eerie precursor of things to come, there also were tales of confrontations between railroad men and boats of Cuban gunrunners picking up caches of arms, and swarthy revolutionary types bound for insurgency operations on the island.

The occasional trip gun, encounter with revolutionaries, or forest fire might have constituted frustration, but any losses were primarily those of time. What truly continued to concern Krome was the possibility of hurricanes. With Flagler having passed another birthday, Krome had already determined that work would continue through the late summer and fall of 1910.

It is a curious phenomenon to be observed among certain hurricane survivors: the area crossed by such a storm is so relatively small and its fury so great that once one has been struck directly and has been fortunate enough to survive, a certain denial, or form of bravado, sets in. *I have been struck by lightning now,* one may be tempted to reason. *What are the chances that it will ever happen again?*

For roulette players, the odds that 00, or any other number, will come up twice in a row are about 1 in 1,200. The odds that a hurricane will pass over the same stretch of land two years in a row are somewhat less. Long odds, certainly, but the fact that the Extension had been battered in 1906 and again in 1909 did not mean that a hurricane could not strike again in 1910.

And so, on the morning of October 17 of that year, as Krome sat at his desk in his Marathon office, reviewing the details of his team's progress and wondering if they could somehow manage to beat the ticking of the Flagler clock, he heard a clamor outside and looked up to find a telegraph operator rushing into his office with a message held aloft, his face an ominous mask.

The clerk was not about to offer his summary of the news, and Krome had to snatch the message from the man's trembling hand. It was a bulletin from the weather office in Miami: a hurricane was approaching, one that was projected to sweep across the Keys sometime that night.

Weather Bureau records indicate that this was an unusual storm, one that had developed a week before over the western tip of Cuba and was presumed to have drifted on into the Gulf of Mexico and died out. As it happened, however, the storm gathered new strength in the Gulf and then drifted backward, striking the same area in Cuba that it had pummeled only a few days before. After this second assault on western Cuba, the storm wobbled northward toward the United States, but was stalled over the Florida Straits by a large high-pressure weather system. Instead of dissipating or spinning away, the storm moved about the warm waters of the Straits, in a tightly contained counterclockwise loop, gaining strength all the while. When the high-pressure system moved off the Florida coast into the Atlantic, this first of the "loop hurricanes" as they came to be called began to bear down on the Keys.

By noon of the seventeenth, winds estimated at 125 miles per hour were pounding Sand Key, where the Weather Bureau's observer reported that "the wharf and woodpile were washed away and the lighthouse shook and swayed in the wind. . . . The force of the wind drew large nails from the doors. The sand was all washed from sight by this time, and monster waves broke over the whole island."

It was hardly welcome news to Krome, but by this stage, preparedness "on the grade" was ratcheted as tightly as he could make it. An agreement had been struck between the FEC and the Weather Bureau, whereby any weather advisory arriving in Washington would reach the camps within ten minutes, and during the hurricane season the most familiar inquiry said to pass between superintendents comparing notes was "How does your barometer read?"

Krome sent the alarm out immediately. Equipment was secured, boats fastened, men moved to shelter in buildings that had been engineered far beyond previous standards.

And yet once again, nature one-upped mankind. The storm of 1906 had concentrated its fury on the Upper Keys, where most of the work ongoing at the time had been centered. The hurricane of 1909 had blasted the Middle Keys, where work had similarly progressed by that time. In 1910, as if guided by an especially malevolent hand, the storm turned its greatest intensity upon the Lower Keys, where unfinished bridge and track work and infrastructure once again made it the most vulnerable segment of the entire Keys Extension.

Even worse, the 1910 storm had slowed greatly by the time it reached shore. The effects of even a catastrophically powered hurricane such as Andrew in 1992, with its 175-mile-per-hour winds, are significantly mitigated if the storm passes over land quickly. Andrew moved ashore just above Homestead, just after midnight, at speeds of up to thirty miles an hour, an extremely rapid rate of advancement. With its storm bands tightly packed, Andrew was gone by 8:00 A.M. And while that 1992 storm proved the most costly in all history, experts pointed out that the toll in damage and lives lost would have been vastly greater had it passed over land at anything resembling a normal rate of speed.

The hurricane of 1910 might have lacked the wind speed of an Andrew, but what it lacked in force, it made up for in staying power. It took more than thirty hours for the storm to pass over the Keys, long enough to convince anyone in its path that the onslaught would never end.

One worker, who held his post on Boot Key as long as he dared, finally called in to Krome to report that his office was knee-deep in rising water. An incredulous Krome ordered the man back to Marathon immediately. When the employee struggled into the main camp, he brought along daunting news: most of the outlying buildings had lost their roofs, and the long wooden trestle that curved off the main line out to the steamship docks had been pounded to smithereens by the wind-driven rains.

Farther south, the damage was even worse. Seventeen miles of roadbed that had been laid across Bahia Honda, Spanish Harbor, Big Pine, and the Ramrod Keys, from MM 37 to MM 20, were washed away. One section of track was found, virtually intact, more than six hundred feet offshore where the winds had dropped it.

One foreman caught in the storm tried to save himself from being swept out to sea by climbing a nearby tree. When the waters continued to rise beneath him and the winds threatened to toss him from his perch in the upper limbs, he tied himself to the trunk with his own belt.

He was still being buffeted by the winds, and at times the tree was bowed down very nearly to the waterline, but the man began to sense that he might live. Then he began to feel a terrible burning sensation at his hands. In moments the searing had moved to his face and lips, pressed tight against the trunk of the tree. His eyes had begun to burn and in moments were nearly swollen shut.

Even in his panic, he realized the terrible irony of what he had done. He had lashed himself to a manchineel tree, one of the most poisonous plants that grow in the tropics, a Lower Keys species of which the men had recently encountered. The indigenous Keys Indians had used the manchineel to poison the wells of the invading Spanish conquistadors, one of whom wrote home to Madrid, "He who sleeps under a manchineel, sleeps forever."

Some workers had been hospitalized after merely brushing against the leaves of the tree; now, as the hurricane raged about him, this man was virtually wallowing in the thick white sap that oozed from its ruptured bark and broken limbs.

Still, what alternative did he have? Where once there had been land, nothing lay about him now but storm-tossed seas.

When the storm had passed, a rescue worker spotted the foreman's swollen body slumped in the branches of the tree, and at first presumed him drowned. Once the man had been cut down, however, it was discovered he was still breathing. Astonished rescuers rushed him to a hospital on the mainland, where his recovery took several months. The bizarre incident constituted one of the grimmer of the human casualties resulting from the storm, which had only one death attributed to its passage.

Perhaps the most troubling discovery that Krome and Coe made in the storm's aftermath was the fact that, in this instance, the sustained winds and waves had conspired to displace one of the principal support piers for the partially constructed Bahia Honda bridge, one that had taken an entire shipload of materials to build. It was an especially disconcerting sign, given the assumptions under which the engineering team had been operating. While wooden trestles and land-laid track might be washed away by storm and flood, such mishaps could occur on railroad right-of-way anywhere, and the destruction could be easily repaired, relatively speaking.

But the thought that one of these massive bridges spanning the channels might be toppled in a storm constituted the worst nightmare of everyone associated with the project. Visions of a fully loaded passenger train toppling into the sea preoccupied Krome and Coe, who went back to work resetting the pier, trying to reassure themselves that the event had been an anomaly.

Still, such uncertainties would linger on, leading to the development of regulations that might seem extreme today. Wind gauges were mounted at the approaches to all the bridges on the Extension, and were connected to electrical safety switches. At any time that the wind speed reached fifty miles per hour, the automatic switches were thrown, and train access to the bridges was denied.

Despite the fact that engineers had calculated that the spans could support speeds of seventy miles per hour and greater, a strict fifteen-mile-per-hour speed limit was imposed on all trains crossing the bridges, even under calm conditions. According to one writer, it meant that crossing the Long Key Viaduct took almost fifteen minutes. Traveling the whole of the Seven Mile Bridge took more than half an hour.

The various setbacks created by the third storm to strike the Extension seemed to take their toll on Flagler, who reeled as though he'd been struck physically by the winds and waves. Though his interest in seeing the project completed did not wane, he began to distance himself from day-to-day oversight of the project, delegating more and more authority to the railroad's general manager, Joseph Parrott, as well as to Krome.

19

Deep Bay

One of the last of the gargantuan tasks involved in the completion of the route was the building of a bridge across the Bahia Honda Channel, connecting Bahia Honda, at MM 37, and the Spanish Harbor Keys. Though the distance was not nearly so great as some of the other spans that had been crossed—a little more than a mile, including approaches—the waters below were the deepest the engineers had encountered anywhere on the route, ranging from twenty feet to as many as thirty-five in some places. The conquistadors had not named the area Bahia Honda (or "Deep Bay") by accident, after all.

Using the rule of thumb that experience had taught—one foot up for every foot of water below—Coe and Krome knew that this bridge had to be built on taller, more massive pilings than any before, in order to keep the rails above the reach of the worst hurricane-driven seas.

The fact that the pilings were so enormous was one reason

why the two men had been so troubled by the discovery that the 1910 hurricane had managed to move one out of place. But, as if spurred on by the realization that Flagler was weakening physically, Coe and Krome went back to work with a vengeance.

Using a combination of the techniques that had been employed on the Long Key Viaduct and the Seven Mile Bridge, Coe deployed nine concrete arches, each eighty feet long, in the shallower waters and connected them with twenty-six longer spans laid atop tall concrete piers. While the top decks of all the other bridges in the vast network were unencumbered by any rails or superstructure, the Bahia Honda bridge was topped by a massive network of support girders more typical of railroad bridges elsewhere, a system in which the tracks are essentially hung from the trusses overhead. (The steel trusses also constituted the only work contracted out by the FEC over the course of the entire project, a decision mandated by economics and by the desire to finish while Flagler was still alive.)

If the result was not as graceful in appearance as the other bridges, this one was certainly the strongest—so much so, in fact, that the builders of the Overseas Highway would one day choose to lay their roadbed *on top of* the Bahia Honda railroad bridge rather than undertake the enormous task of widening it.

Completion of the Bahia Honda Bridge was not the end of Coe's work by any means. Before the railroad could reach Key West, he would build others of his gracefully styled arched bridges, one at Spanish Harbor (seventy-seven spans long) and another farther along at Bow Channel (Cudjoe-Sugarloaf Keys, thirty-two arches), as well as oversee a number of lesser projects. In all, Coe would be responsible for fifteen major bridges on the Extension, as well as twelve miles of permanent trestlework. But the relocation of the displaced pier and the completion of the bridge at Bahia Honda meant that the last of the great challenges had been met and that what had once been the purest of fancies was about to materialize at last.

As early as February of 1911, Krome had been contacted by Joseph Parrott asking a blunt question: "Can you finish the road down to Key West so we can put Mr. Flagler there in his private car over his own rails out of Jacksonville on his next birthday, January 2nd?" While the state law authorizing the project stated that the Key West Extension was to be completed by May of 1912, Parrott's request was based more upon sentiment than upon legal exigencies. Flagler's next birthday would be his eighty-second, and those closest to him had come to feel that the only thing keeping the old man alive was his dream of seeing the project completed.

Krome, who had been a part of things from the beginning, was not unmoved, and agreed to come as close as he could. "I did some close figuring," he wrote, "and finally replied that we could complete the road by January 22nd of that year should no storm overtake us, or no unforeseen delay set us back."

Given the events of the previous six years, Krome's exceptions were not insignificant, but he quickly set about to keep his word. Work schedules, already demanding, were cranked up yet again at both ends of the line. While Coe was overseeing construction of the bridges at Spanish Harbor and Bow Channel, other significant spans were under way at Niles Channel, almost a mile wide, as well as shallow-channel bridges connecting the others of the Lower Keys. Right-of-way in the Saddlebunch Keys, separating Cudjoe and Boca Chica, just a few miles out of Key West, consisted principally of marshland, requiring a return to the laborious dredge-and-fill method of building a roadbed employed just south of Homestead at the project's outset. This time, six of the dredging machines were put to work to carve out a roadbed from the murky shallows.

Still, the question was no longer "Could it be done?" but "Can we make it to Key West on time?"

※   ※   ※

Perhaps the most significant of the remaining tasks had been given not to William Krome or Clarence Coe, but to Joseph Parrott, manager of all Flagler's Florida interests and president of the FEC. Owing to the earlier difficulties with the Navy, there was still no terminal facility built to receive trains or handle the significant shipping traffic that was still using the temporary port facility on Knight's Key.

Flagler dispatched Parrott to Key West, still the most bustling city in the entire state, to survey the situation and recommend how best to proceed. Though Parrott shared Flagler's expeditious nature, this charge was to prove a difficult one.

In contrast to the other East Coast settlements, which had barely been known before Flagler brought his railroad through at the end of the nineteenth century, Key West had been a thriving port and commercial center since the 1820s. From the time of the first seventeenth-century explorers, the island had been a haven for pirates, who operated largely without interference until the U.S. Navy drove them out in 1822 and the city was chartered.

After the pirates, the next group of entrepreneurs to come along was the wreckers—or *gentleman* pirates, depending on one's point of view. These wreckers built observation towers on land from which they kept a close eye on the outlying shallow reefs, especially during storms. When they spotted a ship that had run aground, they sped to the scene, often warring with their competitors on the way. Whoever made it to the site first was usually good enough to take survivors on board, but according to the law of the sea, the cargo belonged to the first to pull it from the wreck.

A favorite local story, and one retold by longtime Key West resident John Hersey in his *Key West Tales,* concerns the exploits of a Methodist preacher known as Squire Egan, who, during the week, plied the nearby reefs as a wrecker aboard his swift schooner *Godspeed.* On one particular Sunday the squire regaled his flock with a sermon based on the text of I Corinthians 9:4: "Know ye not that they which run in a race run all, but one receiveth the prize?" Hardly had he begun than Squire Egan glanced out the window from his high pulpit to catch sight of a ship running aground on the nearby reef.

The squire did not miss a beat, however, and continued his sermon as he stepped from the pulpit and walked down the aisle between his rapt parishioners, leaning close and admonishing row after row to prepare themselves for the great race toward salvation.

When he had reached the great doors at the back of the church, he turned to give the congregation his benediction: "Wreck ashore," he called. "Now we will run a race and see who receiveth the prize. Run that ye may obtain." And with that he was out the door toward the *Godspeed,* all the men of his congregation in close pursuit.

The story may be apocryphal, but the practice was certainly not. Because the reefs were unreliably charted and often shifted position with the tides, the wrecking business was lucrative, so much so that it was not unknown for a captain to arrange a convenient wreck in advance. Once the cargo had been claimed and sold, the proceeds would be divided between the captain and the wreckers, with the owners left to fight it out with insurers, if they had any.

Matters had reached such disarray that a federal court was established in Key West in 1822 to try to bring order to the situation. Court or no, when storms abated and things got slow, some of these angels of mercy would go so far as to erect false lighthouse beacons that were sure to lure unsuspecting vessels onto the rocks. An unfortunate captain might complain, but by the time authorities made it out to check, all traces of any bogus light had vanished.

A number of those shipwreck survivors were introduced to the beauties of the atoll-like paradise in such a manner, and many of them stayed on to make a life there. The bounteous local waters supported thriving industries in fishing, shrimping, sponge diving, and even turtle raising.

As the word spread, others flocked southward. Even John James Audubon was a frequent visitor, drawn by the seemingly limitless flocks of sea and shore birds in the area. Lest anyone assume that the famed ornithologist was among the original tree-hugging conservationists, however, local historians are quick to point out that he should be more properly known as the Buffalo Bill of the bird population. Audubon and his traveling companions were reported to return from bird-hunting forays with hundreds and sometimes thousands of specimens in their bags.

It was a practice that Audubon freely defended—how else to study the creatures at close quarters, or to find that perfect specimen to serve as a model for his work? But he was also a thrill-seeking sportsman who considered it a bad day when he had downed fewer than a hundred birds.

"We have drawn seventeen different species since our arrival in Florida," Audubon once wrote, "but the species are now exhausted and therefore I will push off."

If Audubon was pushing off, however, thousands more were pushing in. By the 1880s, Key West had become the center of a thriving cigar-making business, much of it presided over by immigrants from nearby Cuba, fleeing the iron-fisted rule of Spanish colonialists. At the industry's height, and before most such work moved northward to Tampa, Key West rollers were producing more than 62 million cigars a year, according to *Tobacco Leaf,* an industry journal. Others put the figure as high as 100 million.

By 1890, then, Key West was the most populous city in Florida, with its port ranked as the thirteenth busiest in the nation and its one by four miles of territory built and overbuilt already. By the time Joseph Parrott arrived, there was simply no more land to be had, certainly not enough for Flagler's ambitious plans.

"There is no more dry land in Key West," Parrott reported to his boss.

"Then make some," Flagler replied.

And Parrott did.

He hired Howard Trumbo to head up the completion of the terminal project, explaining something of the urgency behind their efforts, and Trumbo went to work with the gusto befitting a member of Flagler's team. He constructed a bulkhead extending in a broad arc above the northwest corner of the island and dredged thousands of cubic yards of marl from the adjoining flats to fill in a breakwater and foundation for railyards, terminal buildings, and docks. Part of the complex included a 1,700-foot-long pier, wide enough, at 134 feet, to allow trains to pull directly alongside a docked ocean liner.

Once again the Navy tried to block the project, complaining that Trumbo was removing fill from submerged lands under their control, fill that might be needed for defense purposes someday. Parrott's response was couched in classic Flaglerese: If the time ever came when the Navy needed its mud, Parrott said, they had his word it would be returned from whence it came.

Such a promise probably did not satisfy Navy officials, but with Flagler's railroad about to reach the island, excitement among the local citizenry and officials was at a fever pitch. Any wrangling over mud and dredging issues was moot. Trumbo's work would go forward, and by January of 1912, the terminal facility at Man-of-War Harbor was in readiness: arriving passengers would be able to walk a few steps across a platform and board a ship bound for Havana. As many as five hundred freight cars could be stored in the yards, ready to load exotic cargo brought up from the Caribbean and Central and South America, then hauled directly north to, as Jefferson Browne had once predicted, "provide the country with pineapples, tropical fruit and vegetables all winter."

By this time even Flagler's former enemies had been swept up in the mounting enthusiasm. Representative Frank Clark, a Florida congressman who had staunchly opposed Flagler at the time he sought to have Florida's divorce laws changed, had come to an about-face. In a speech before the Florida House, Clark lauded Flagler's contributions to the state as having been "the direct cause of providing happy and contented homes for full 50,000 people," and went on to introduce a resolution celebrating the project's impending completion and applauding Flagler's vision and endurance.

Local editorial writers were jubilant, predicting a limitless future of prosperity for the island paradise. "The Old Key West—one of the most unique of the world's historic little cities—is shaking off its lethargy," said the *Florida Times Union,* ". . . the spirit of progress and development will be greater than ever." A new year and a new era were about to dawn.

20

Wonder to Behold

On the afternoon of January 21, 1912, almost seven years after work on the Key West Extension had begun, the project's equivalent to the driving of the "golden spike" took place. At Knight's Key, nearly fifty miles north and east of Key West, a bridge foreman threw a switch that closed off access to the trestle curving from the main line toward the temporary docks. For the first time, traffic was open across the Seven Mile Bridge—at the time, the world's longest continuous bridge—and from there, all the way to Key West. The process of rail building that had begun in 1892 was complete. There were now 366 miles of FEC track linking Jacksonville with Miami, and 156 more connecting Miami with Key West.

That same morning Henry Flagler, now eighty-two, left his home, Whitehall, in Palm Beach. He was frail and his sight was failing, but nothing was about to stop him. Not after spending $12 million on a series of hotels, $18 million on his land-based railroad, and another $20 million or more on his "railroad across the sea." On this day he would board his private railroad car at the West Palm Beach station for a 220-mile trip that would culminate in Key West and punctuate the dream of a lifetime.

Typically, the journey was not without its overtones of threat. While all rails and bridges were at last in place, and hurricane season was well past, this project would not end without another obstacle arising: the FEC's firemen had picked this time to go on strike. As a precaution against sabotage in the notoriously volatile arena of railway labor dealings, shotgun-toting security men were posted along every mile of the route.

Despite the company's fears, the first leg of the journey went without incident, and Flagler spent the night in Miami, a stopover planned so that he could arrive in Key West early the following day, fresh and rested. The *Miami Herald* reported that the *Extension Special* left the city that morning under especially brilliant skies, as if ordered by the fates.

Attached to the engine were five cars of FEC officials and invited dignitaries, the last being Flagler's own sleeper, Car 91, which he had used to travel the rails since having it built in 1886. Car 91 was a copper-roofed pleasure palace of a railroad car, containing a Victorian-styled, wood-paneled lounge, sleeping berths for visitors, and a private stateroom with bath for Flagler. There was a copper-lined shower, a dining area, and a small food preparation area with an icebox and a wood stove.

Among the various guests on board with Flagler on the morning of January 22 was Assistant Secretary of War Robert Shaw Oliver, the personal representative of President Taft, sharing a ride that was so sought after that it took a last-minute call from William Krome to Flagler direct to get himself on board.

Among the ten thousand or more assembled in Key West to greet Flagler were emissaries of President Gómez of Cuba and of the governments of Italy and Portugal, as well as numerous Central and South American countries. While the *Extension Special* had actually been preceded down the line on the previous day by a scout train sent to test the rails, no one mistook the significance of this morning's ceremony.

"[T]he railroad magnate has extended his rod, the sea has been divided," wrote a reporter for the *Key West Florida Sun*.

In his volume *Key West: The Old and the New*, published the day of the ceremony, longtime proponent Jefferson Browne said, "Henry M. Flagler's railroad . . . is man's last word on that marvelous style of construction and will echo through the ages to come," adding, "Everything that went into the construction of this work obeyed his will. . . . The Greeks before Troy suffered no greater hardships, no greater heroism."

And at 10:34 A.M., Henry Morrison Flagler, his back bent with age and his dim eyes brimming with tears, stepped out onto Car 91's observation platform to an ovation the likes of which he had never encountered. He had "ridden his own iron" to Key West at last.

The mayor of Key West welcomed him with a fulsome speech and, on behalf of the citizenry, more than half of whom were present, presented Flagler with a gold and silver commemorative plaque bearing his likeness. Another golden tablet came "To Uncle Henry" from the men who had labored on the project those seven long years.

A military band played, and a children's chorus of one thousand voices sang patriotic songs in Flagler's honor. A choked-up Flagler turned to Parrott and whispered, "I can hear the children, but I cannot see them." Parrott, nearly overcome himself, simply gripped his old friend's arm and squeezed.

When finally called upon to speak, Flagler managed to rally. "We have been trying to anchor Key West to the mainland," he said, ". . . and anchor it we have done."

Then, gazing out over the vast assembly, he uttered what for Flagler was an outright effusion: "I thank God that from the summit I can look back over the twenty-five or twenty-six years since I became interested in Florida with intense satisfaction at the results that have followed."

No one will ever know just what thoughts ran through Flagler's mind as he stood before the crowds that day, of course. So many years had passed between the undertaking of the project and its completion that some sense of anticlimax would have been inevitable. After all, he had known for months now that the final rails would one day be laid and that the project that so many had turned away from and others had derided would become a reality.

THE EIGHTH WONDER OF THE WORLD, headlines now bannered, such epithets as "Flagler's Folly" long forgotten. He had arrayed before him thousands of grateful citizens, along with a multitude of foreign dignitaries and government officials come to pay homage to what had been accomplished solely because of his vision and his unswerving devotion to that objective. Few people in history have accomplished so great a task or lived to experience such a moment as Flagler did.

The man he'd hired to bring his dreams to fruition had died on the job and hundreds of other men had lost their lives as well, and despite all bromides otherwise, some weight of their passing had to have rested upon Flagler's shoulders. Storms weathered, court fights fought, political enemies bested, impossible engineering problems solved, good men buried, rails joined at last. So many currents, so many thoughts and notions to meld and comprehend, after eighty-two years of life.

There's no way to fathom how much of this had passed through his mind that day, but on his way off the platform Flagler placed a hand on Parrott's shoulder and whispered, "Now I can die happy. My dream is fulfilled."

The celebration in Key West went on through the day and evening, and lasted for days. An elaborate series of events posted in the official souvenir program written by George M. Chapin for the company included aviation meets, boat tours of the island and the railroad facilities, yacht races, five nights of fireworks displays, and more. An opera company from Spain gave a series of performances, as did a circus—"Publillone's Great"—brought over from Cuba. A carnival was set up on the terminal grounds, and even the U.S. Navy had assembled an honoring flotilla of warships in the nearby harbor.

Several other trains carrying fare-paying passengers followed the *Extension Special* into Key West that day, including one advertised as the "New York to Havana Special," which, company literature boasted, constituted the longest sleeping-car passage in the eastern United States. Passengers on that train were escorted directly from the platform to the *Governor Cobb,* waiting at dockside, which sailed for Cuba that same afternoon.

At a ball that evening, Florida's governor, Albert Gilchrist, lauded Flagler yet again, issuing a proclamation that stated, "The building of this great oversea railroad is of nationwide importance, second in importance only to the construction of the Panama Canal." President William H. Taft sent along a personal note of congratulations. But by this time Flagler was exhausted from the round of appearances and activities. The next day he boarded Car 91 for a return to his home in Palm Beach.

Given a bit of time to reflect on the events of the past few days, Flagler sat down on January 27 to compose a letter to Joseph Parrott, one that gives some insight into his state of mind:

The last few days have been full of happiness to me, made so by the expression of appreciation of the people for the work I have done in Florida. A large part of this happiness is due to the gift of the employees of the Florida East Coast Railway. . . . I beg you will express to them my most sincere thanks. I greatly regret that I cannot do it to each one in person.

The work I have been doing for many years has been largely prompted by a desire to help my fellow-men, and I hope you will let every employee of the Company know that I thank him for the gift, the spirit that prompted it, and for the sentiment therein expressed.

It was as much as Flagler would say about his feelings, and whatever lay in his heart of hearts matters less today than the legacy of his works themselves. It is the lesson of all heroic tales, after all. Not the doer, but the deed. Not the man, but all mankind.

21

Failed

I f Henry Flagler was in fact motivated by visions of profit in building the Key West Extension, and if in fact profit is to be the rule by which that mighty effort is measured, then those who scoffed and called the endeavor folly from the outset were eventually proven correct. It is true that the Key West Extension became an instant hit with passengers, but freight traffic, which has always been the backbone of the railroad business, never materialized to the degree that Flagler had predicted.

He had invested somewhere between $27 million and $30 million to lay its 156 miles of track, more than half again what it had cost to run the railroad 350 miles down the mainland coast of Florida, all the way from Jacksonville to Miami. And even mainland operations, which enjoyed relatively busy freight traffic, were limping along, losing as much as $400,000 a year.

Expectations of a flood of imported goods to be hauled up

from the Caribbean and Central and South America did not materialize, for the simple reason that there was very little to import, aside from what was already making its way to the United States through existing channels. It was a realization similar to that made by U.S. entrepreneurs following the end of the Cold War and the dissolution of the Iron Curtain. Projections of a massive burgeoning of trade between the West and the Communist bloc came to little, for those newly opened nations were producing virtually nothing of interest to highly developed nations and lacked the capital to import Western goods or create the technical or industrial infrastructure required for capitalist enterprise. Even in this modern age, the effort to jump-start such reciprocity stumbles and wheezes along, well into its second decade.

In addition, while Parrott and Trumbo had managed to get in place a docking facility suitable for maintaining passenger trade between Key West and Havana, Flagler's dreams of a giant deep-water port with twelve massive covered piers eight hundred feet long and two hundred feet wide would never come to fruition. Without major docking and freight transfer facilities, steamers, then and after the opening of the Panama Canal, continued northward to New Orleans, Tampa, Jacksonville, and elsewhere.

The company did its best to increase freight traffic, though, by building three huge ferries constructed especially to haul freight cars. These 360-foot ships—the *Henry M. Flagler,* the *Joseph R. Parrott,* and the *Estrada Palma*—were the largest train ferries in the world at that time, and were fitted with standard-gauge rails in their holds. The ferries could back directly to the docks, where fully loaded freight cars could be shunted directly on board, as many as thirty-five at a time. It was a practice that Flagler himself had visualized, ever since his discussions with Canadian railroad baron Sir William Van Horne. If Van Horne's ferries could transport rail cars over 112 miles of rough seas on the Great Lakes, then, Flagler reasoned, he could surely do the same on the hundred-mile passage between Key West and Havana.

The principal cargo of the ships proved to be pineapples, most of them brought up from Cuba during the six- to eight-week season. Thirty-five hundred carloads of the fruit passed through Key West each year, and unemployed workers enjoyed a bonanza each "Pineapple Day" when a new shipment arrived. Cars coming across the straits from Cuba were always greatly overpacked to save space and shipping costs, and the relatively fragile fruit had to be sorted into new containers before being sent along to its ultimate destination.

Going in the other direction, the cargo often consisted of U.S.-bred hogs, their pork considered a great delicacy in Cuba. The aroma that wafted from the resultant stockyards on Trumbo Island lent an entirely new aspect to close-quarters life on Key West, with nearby residents able to reckon what kind of day it was going to be by the direction the breeze intended to take.

Passenger traffic was another matter altogether, with a direct link between New York and Havana as amazing and enticing to travelers then as the opening of the London-to-Paris "Chunnel" became nearly one hundred years later. Soon after the *Extension Special* rolled into Key West, regular service to Havana began. As a Havana newspaper reported, "This train carried the latest design all-steel Pullman drawing room and Standard sleeping cars, is electrically lighted and equipped with electrical fans throughout. There is no change of cars between Key West and the Pennsylvania Station in the very heart of New York City."

The schedule boasted that passengers could leave Havana at 10:30 A.M. daily, except Sundays, and arrive in New York at 7:55 P.M. on the second evening following, though, owing to weather and other delays, it was a promise more often honored in the breach. The Key West–Miami run was listed as requiring only four and one-half hours, not much more than it takes to drive today, but locals who rode the route often reported the journey required the better part of six or seven hours.

One retired office engineer for the FEC divulged something of the truth of the railroad's policy. "Whenever a train was twenty-four hours late, it was never admitted," he told an interviewer. "The bulletin board would read 'One Hour Late,' failing to state it was one day *and* one hour late."

But in the tropics, delays are relative. Vacation travelers from the frigid Northeast hardly minded a train puttering across a bridge at fifteen miles an hour if it meant they could stare down from their windows at schools of dolphin rolling alongside through the waves. And even the prospect of sitting before the gate of a closed bridge, waiting and watching as a line of glowering thunderheads swept across a white-capped channel, was not without its drama.

As a writer of the day enthused, passengers could "watch the stately procession of southbound ocean steamers which pass . . . in the great tide of traffic to the West Indies, and to Central American and South American ports. Nor is it at all fanciful to suppose that if he is wise enough to carry along a fishing line and bait, he may find sport from the car platform should the train happen to halt on the Long Key or Bahia Honda viaduct."

To those who had basked in Flagler resorts in Ormond Beach, St. Augustine, Daytona Beach, Palm Beach, and Miami, a trip along the Key West Extension with a few nights' stopover at Long Key, a few more in Key West, topped off with a foray to Havana, was an undertaking that could countenance the languid approach and the demands of life in the isolated Keys. The *Havana Special* indeed operated as an express all the way from New York City to Miami, but from there southward, all bets were off, with the trains making as many as forty-three stops on the way to Key West.

Once rail service was established, the U.S. Post Office discontinued the use of boats to carry mail to the Keys, and every station down the chain had its postal operations. Other shippers followed suit, with regular stops required to off-load hard-to-come-by staples to Conchs and hardy settlers along the line.

One family living on Big Torch Key, far down the chain, had a Model T Ford truck delivered to the station at Ramrod. There was no way to get the truck across the channel separating Ramrod from their homestead on Big Torch. They used it solely to pick up what could now become regular deliveries of that fabulous substance known as ice. The homesteaders would load the big blocks onto the truckbed, then drive it down a road they'd hacked through the wilderness to a dock they'd fashioned, then transfer the load to a skiff to transport it to their home. Such were the small miracles made possible by the railroad across the sea.

But the ice runs to Ramrod and the passenger traffic on the *Havana Special* were scarcely the stuff of which railroad profits were made. In fact, there is little evidence that the FEC system as a whole was making much money, from its rail lines, at least. Associated land development, hotel operations, newspapers, and other Flagler enterprises in Florida did well enough, but it was nothing to compare with what Standard Oil had produced.

One Flagler refrain—"Go to Key West—" was to be replaced by another: "I would have been a rich man if it hadn't been for Florida."

Meanwhile, Flagler continued to work, traveling from his Palm Beach home to spend the summer at Satan's Toe, near Mamaroneck on Long Island Sound, from where, despite his various infirmities, he commuted regularly to company offices in Manhattan. If he was disappointed in the fiscal performance of the Key West Extension, he wasn't saying so publicly. In the fall, he and Mary Lily traveled back to Palm Beach, and on January 2, 1913, Flagler celebrated his eighty-third birthday at Whitehall.

Normally, he and his wife would have moved up the coast to St. Augustine by late February, where they would stay until the warm weather encroached. In mid-May, it was their practice to board Car 91, or sometimes one of Flagler's yachts, to travel back to Mamaroneck. But this year the Flaglers prolonged their stay at Whitehall, closing down most of the enormous house and sending the staff away.

According to an account rendered by David Chandler, one of Flagler's biographers, on the morning of March 15, Flagler went to use a bathroom on the main floor of the house normally reserved for staff and guests. The door to the bathroom opened outward, giving access to a narrow threshold that stood at the top of a short flight of stairs leading down to the bathroom itself. In keeping with Flagler's fondness for the latest in conveniences, the door had been fitted with a powerful pneumatic device that would pull the door firmly shut behind him.

No one knows exactly what happened, of course, but it is possible to speculate:

Flagler, alone in this part of the cavernous mansion, moves gingerly inside the doorway of the guest bathroom and pauses at the top of the stairs. Perhaps he fumbles for the light switch on the wall just inside the door and to his right. Perhaps he feels a moment's light-headedness and intends to call for help.

But, whether he has turned for help or is simply poised there for a moment before descending, the powerful pneumatic device Flagler has installed does its appointed duty, a bit too swiftly and surely.

The door whisks shut. And in its path is a frail and fragile man of eighty-three.

Like a tidal surge blasting through a channel in the Upper Keys, the rush of the heavy door is an implacable force. Flagler finds himself thrown forward, his feet flying from the top of the staircase into the sudden void. And then there is only darkness.

      ⊠   ⊠   ⊠

Chandler attributes this theory to Charles Simmons, a former director of the Flagler Museum, located in the building that was once Flagler's opulent Palm Beach home, Whitehall. But current museum curator Sandra Barghini watches the same door swing shut, propelled by its powerful pneumatic closer, and simply shrugs. "Doctors say most old people break their hips just by walking," Barghini says. "And *then* they fall."

Regardless of just how it happened, it was not until several hours later that one of the few remaining servants discovered Henry Flagler sprawled unconscious at the bottom of the stairs, his hip fractured, his body battered and bruised. An orthopedic surgeon who happened to be staying at the nearby Poinciana Hotel was summoned and Flagler was revived. He was removed to a makeshift hospital room arranged in the Nautilus, one of the hotel's beachside cottages that Flagler often used. There he lingered for two months, immobilized in bed by heavy sandbags and disoriented by painkilling drugs. Though he rallied briefly at times, it became clear that Flagler would never leave his bed again.

On the morning of May 20, 1913, with Mary Lily at his bedside, Henry Morrison Flagler died. His body was taken to St. Augustine for burial, after a simple service conducted in the Memorial Presbyterian Church, which he had built. Flagler's pallbearers included Joseph Parrott, William Krome, James Ingraham, and William Beardsley, all men who had been with him through the immense task of the Florida East Coast Railway to Key West.

Duly noted: John D. Rockefeller did not attend Flagler's funeral. There is no record that any member of the far-flung empire of Standard Oil did so.

22

Rolling On

Ⓞne magazine writer of the day suggested that the completion of the railroad would mean the end of life as it had been known in the Keys:

> *And now their strange inhabitants, their white and shimmering silences broken only by the cries of gulls and the long roll of blue waves breaking on coral rock are to know the shriek of the locomotive and the roar of the passing trains. . . .*
>
> *The novelists will have to move over to the West Indian islands or across the Caribbean to find homes for their smugglers, absconding cashiers, and the lone lovely daughter of the irascible, civilization-hating hermit. The whistle of the locomotive will be heard in the land and another queer corner of the earth will be put on the civilized map.*

A dire prophecy indeed. But while the railroad did affect life in the Keys, it was not in the cosmic fashion suggested. That "queer corner of the earth" proved too tough, for one thing; and, for another, economics proved to be a science nearly as difficult to forecast as hurricanes.

◦   ◦   ◦

Historians have made much of the fact that Flagler did not live to see the downturn in the fortunes of his Key West Extension, but that view seems to give the man little credit. Flagler was one of the cofounders of the most successful corporation in the world, after all, and was certainly savvy enough to know that despite all his stiff-upper-lip talk about controlling shipping traffic on the eastern seaboard and opening up whole new continents to trade, he could have invested the $20 million to $40 million expended on the extension in far more profitable ways. And certainly, over the year and more that passed following the opening of the Extension, he had ample opportunity to assess the direction that things were taking.

In July of 1912, for instance, William Krome passed along the disturbing news that a case of bubonic plague had been confirmed in Havana. Krome feared that the disease might be carried back to the United States by passengers on one of the company's ships and wrote to FEC physician J. M. Jackson in Miami, seeking advice on how to detect signs of the illness and prevent its spread. Plague, however, was one scourge that the Key West Extension was spared.

Freight, a certain amount of it, was carried up and down the Keys, as were droves of awestruck passengers. The railroad's presence also hastened the development of island communities that would likely have languished as forlorn outposts for decades, though whether that is necessarily a good thing remains an issue of hot debate in a region where natural resources are limited and the ecosystem is so fragile.

After the Volstead Act was passed in 1920, the *Havana Special* became a favorite of the sporting set. Key West then, as now, did not regard itself as fully beholden to a system of laws and regulations developed without input from the "Conch Republic." The city's bars were never closed, the rumrunners never stopped their trade between Key West and Cuba, and federal agents were known to beg off any posting to the wide-open "Southernmost City."

In an interview with railroad historian Pat Parks, whose monograph *The Railroad That Died at Sea* was the first to tell the Extension's story, former FEC baggage master Kingman Curry remembers many a massive steamer trunk labeled "Wearing Apparel" gurgling and sloshing merrily as he trundled it from ship to railroad car. According to Curry, an astonishing number of passengers who made the trip down to Cuba during Prohibition seemed to have died there.

"I never saw a death certificate on one of those coffins that gave any other cause of death but 'alcoholism,' " Curry told Parks. "It's possible some of them contained demijohns of rum. But it wasn't my duty to open them. Trunks and coffins were all sealed by customs inspectors."

The Casa Marina Hotel, which Flagler had always intended to be the crown jewel in his chain of pleasure palaces, finally opened its doors in 1922. The fortresslike building had walls nearly two feet thick at their base, and with its private beach, tennis courts, and lush landscaping, became an immediate hit with well-to-do travelers. Despite stints as a Navy barracks during World War II and an Army staging base during the 1962 Cuban missile crisis, the hotel has been restored to its original state and still lures the rich and famous through its doors.

Despite the worst fears of the railroad's builders, no storm-battered passenger train ever plunged off one of the mighty bridges to disappear in a hissing sea, though there were a few minor collisions and accidents reported. One of the worst took place in 1927, when a freight engineer, apparently incapacitated by the flu, took his 110-car train up the line from Key West without bothering to check the water level in his boiler. As the train was laboring up the grade at Seven Mile Bridge, the overheated boiler exploded, raining pieces of the engine chassis into the waters. The unfortunate engineer was killed and his fireman badly injured, but no one else was hurt.

As for predictions that Key West, "America's Gibraltar," would become a burgeoning center of commerce, Trumbo's difficulties in finding space for so much as one steamship dock should have been a tip-off. Even before the railroad came miraculously to town, Key West was decidedly overbuilt, and the infrastructure for expansion simply did not exist. There was no ready supply of fresh water, no highway in or out, and only a single railway line to service the projected flood of trade.

Instead of encouraging growth, there is evidence that the railroad's arrival actually encouraged some Key West residents to leave the far-flung island. At last, significant numbers of immigrants who had come from Cuba and other Caribbean islands in search of a better life had ready access to a larger world—a Sunday round-trip ticket to Miami went for $2.50, and many who made it that far north simply made up their minds to stay on the mainland. As effects of the Great Depression swept as far southward as Key West, the trickle of those moving northward became a flood. By 1930, the U.S. Census revealed that Key West had actually *lost* more than seven thousand residents.

Other troubles followed in the wake of Flagler's passing: his trusted manager Joseph Parrott, who took over direction of the company upon its founder's death, would himself die scarcely five months later, of heart failure, amid growing complaints from pioneering growers up and down the Keys that they were being forced out of business by the importation of cheaper pineapples and tropical fruits being hauled up the FEC line from Cuba.

And Mary Lily's life was hardly charmed thereafter. From 1913 on, she divided her time between Whitehall and New York, and in 1916 married an old friend, Robert Bingham, then an ambitious but struggling attorney in Louisville. Scarcely eight months later, Mary Lily died, shortly after amending a will that left her new husband $5 million of the $100 million Flagler's estate had bequeathed her.

While fifty-year-old Mary Lily's death was reported as heart failure, the timing led to whispers of her acute alcoholism and even laudanum abuse, a not-unheard-of practice among certain of the leisure class of the day. The fact that Bingham used his inheritance to pay off a significant number of debts and its remainder to buy the *Louisville Courier-Journal* and the *Louisville Times* fueled rumors that if Bingham hadn't actually been involved in Mary Lily's death, he did little to prevent it.

Writer David Chandler passed off Mary Lily's death as an "apparent heart attack" in his Flagler biography of 1986 (subtitled "The Astonishing Life and Times of the Visionary Robber Baron Who Founded Florida"). But by 1987, he had changed his mind. *The Binghams of Louisville*, a biography coauthored with his wife Mary Voelz Chandler, argued that Bingham had clearly aided and abetted Mary Lily's death. By 1991 three more Bingham family retrospectives were published, with the last in the series, Susan Tifft and Alex Jones's

The Patriarch, going so far as to charge that Mary Lily had died of cardiovascular syphilis, a disease most likely contracted from Henry Flagler himself.

Subsequent historians have dismissed such speculations as pure sensationalism, but it is a reminder that if Flagler truly had immortality in mind as he pressed ahead with the fight to complete the Key West Extension, he should have understood the costs of such status.

Even charges against the railroad sometimes grew to near-outlandish levels. In 1927 the British Isles, along with the rest of northeastern Europe, suffered through a particularly severe winter and subsequent summer marked by record low temperatures—which, along with powerful storms, resulted in deaths, property damage, and crop failures. Pre–El Niño–era scientists seeking some plausible explanation for the calamities pointed fingers all the way to the Florida Keys, where, it was suggested, the embankments and the bridges of the Key West Extension had caused a permanent divergence in the flow of the Gulf Stream, the steady flow of warm water that originates in the Caribbean and runs all the way to the coasts of France and England, greatly ameliorating what would otherwise be forbidding climates.

Despite their extreme nature, the complaints grew so shrill that the U.S. government was forced to commission a study to investigate the matter. No evidence was found that the Gulf Stream had been displaced, however, and by the following winter, things had returned to normal in the European weather pattern.

❧   ❧   ❧

Part of the reason for various speculations about Flagler is undoubtedly his close-to-the-vest nature. As far as anyone knows, even the most traumatic events in his life produced no soul-searching confidences or outpourings of grief. Certainly no bundle of letters or heartfelt diary entries have surfaced to counter the image of a supremely stoic man.

A story was told by one of Flagler's subordinates, who, having learned of a stunning reversal in the fortunes of Standard Oil, rushed into Flagler's office to deliver the good news that a judge's fine of $29 million levied earlier against the company had been reversed. Flagler had been greatly upset by levy of the fine; not only was the amount unprecedented, but he considered it a direct result of Roosevelt's personal vendetta against him.

And yet, as his employee stood there, waiting for some momentous reaction, there was only this:

"For a moment, he looked as if he were going to say something. But he merely nodded and then said casually: 'Mr. _____, do you happen to have those Whitehall plumbing bills handy?' "

Just as typical are the terse entries to be found in his 1909 diary, a year of many milestones for the Extension project:

For Saturday, January 2, 1909, Flagler's birthday: "Began my 80th year. HMF"

For April 20, 1909, the day he lost his project supervisor: "J. C. Meredith died at 3 P.M. in the hospital in Miami."

For August 20, 1909, the momentous beginning of construction on the "impossible" Seven Mile Bridge: "First steel span on Knight's Key Bridge was put in place at 3 P.M."

For October 11, 1909, after receiving the devastating news of the previous day's storm: "Hurricane passed over Florida causing great damage at Key West and on Key West extension. Tug Sybill and 13 of her crew drowned. Also Brown at Marathon."

By 1910, Flagler had dispensed with any mention of his birthday. And as for the impact of the hurricane of 1910 on the Extension project or its leader, not a word was set down.

As Edwin Lefevre wrote in his 1910 piece for *Everybody's*, "Flagler's is not only an excessive modesty but a personality so elusive as to be unseizable. . . . He has no intimates."

Toward the end of his time with Flagler, Lefevre admits to growing frustration with his subject:

"You don't seem to care to talk about yourself," he said to Flagler.

"I prefer to let what I have done speak for me," Flagler replied.

"By their works ye shall know them," Lefevre suggested.

"Yes; that's it," Flagler said—as eagerly as he had said anything, according to his interviewer.

During their final encounter, Lefevre made one last attempt to get to the core of the man who would have been the second richest in the world after his former partner John D. Rockefeller, had it not been for those Florida investments.

"All that day I had tried to catch a glimpse of this man's soul—in vain . . . and now in the loggia of his palace, looking out to Lake Worth . . . I turned . . . and . . . I asked him, I fear impatiently:

'Doesn't this sky get into your soul? Doesn't that glow light it . . . isn't *this* the real reason why you do things here?' "

Flagler seemed to think about the question, says Lefevre. Then he turned slowly and said, "Sometimes, at the close of day, when I am fortunate enough to be alone, I come here. . . . I look at the water and at the trees yonder and at the sunset . . . [and] I often wonder if there is anything in the other world so beautiful as this."

It was about as much of a straight answer to a personal question as anyone was to get from Henry Flagler.

⊠   ⊠   ⊠

By the time the Panama Canal was opened in 1914, much had changed in Key West. Most of the cigar factories had moved northward to Tampa, lured by subsidies dangled by local business leaders. A peaceful Caribbean political situation had led to a substantial decrease in size of the Navy base. And the conversion of many oceangoing steamships from coal to oil as a fuel meant that longer runs were possible without refueling.

The enactment of tough new trade tariffs would greatly reduce the imports of Cuban tobacco, sugar, and pineapples. New taxes levied on the FEC drove up prices, sending business to competitors sailing from New Orleans and elsewhere.

And yet, even without its visionary at the helm, the Key West Extension managed to roll on, struggling through good economic times and bad, occasionally assailed by critics, yet always a favorite of travelers drawn to the American tropics, to Key West, and beyond. By the 1930s a round-trip ticket from Miami to Havana cost as little as twenty-four dollars, and even through the early years of the Depression, daily service continued on the famed *Havana Special*.

The appeal of the Keys as an exotic destination had only been bolstered by their newfound accessibility, so much so that Franklin Roosevelt had thought it a worthy expenditure of WPA funds to complete a highway link between Grassy Key and the Matecumbes, thereby making it even easier for Americans to find a part of paradise for themselves. What effect that highway might have had on an already struggling rail line was a question debated by railroad officials and Middle Keys residents who were somewhat divided on the prospect of a second artery of travel through their midst.

It was an intriguing question, of course, but it would never be answered. Before men could finally address the issue, nature decided the matter by herself.

23

Storm of Storms

Hurricanes, scourge of the Key West Extension during its construction, seemed to lose interest once the line was in place, for a good long while, at least. Even the monster blow of 1926, which struck Miami head-on and virtually destroyed the spanking new suburb of Coral Gables, laid hardly a scratch on the Keys Extension. In one of history's ironies, George Merrick, another notable Florida visionary and the founder of the city of Coral Gables, was forced into bankruptcy by the 1926 storm, and only managed to survive by cobbling together enough backing to open a fishing camp on—where else—Matecumbe Key, not far from where Flagler's Long Key Fishing Camp was still operating successfully.

Other hurricanes had pummeled South Florida over the twenty-five years since the last to strike the Keys in 1910, some of them with profound consequences. But if the Key West Extension wasn't making money, it seemed to have entered a charmed phase of existence, in one regard at least. The line had taken three major strikes during the time it was being built. A person might have been forgiven, then, for thinking that as far as danger from hurricanes went, the railroad across the ocean was out.

A fanciful notion, possibly a comforting one. But hardly a compelling one. Certainly, railroad men who had been around from 1906 to 1910 wouldn't have bought into it—and a pity there were few of those left on duty in 1935. Joseph Parrott had died shortly after Flagler, and William Krome, who had retired to a farm in Homestead following Parrott's death, had passed away in 1932. Clarence Coe, the masterful bridge builder, was still alive, but he was long out of railroad work. He'd gone on from the FEC to become the first city manager of Miami, and by 1935 he was serving as chief of that city's Public Housing Authority.

⌗   ⌗   ⌗

It is interesting to speculate whether Krome or Coe might have done things differently, had they been involved in the railroad chain of command that fateful Labor Day weekend of 1935. It is possible they might have made a difference, had they known the power of the hurricane that was bearing down upon the Keys—had they been able to foresee Hemingway lurching through the fringes of the storm toward his boat; Bernard Russell watching his sister and her baby being peeled from his arms by the winds; workmen seeing friends impaled by flying timbers and decapitated by flying sheets of tin; railroad engineer J. J. Haycraft staring in disbelief at a wall of water about to swallow his train, along with the very world.

But little had changed in the science of hurricane forecasting between 1910 and 1935, and without the monster staring you in the eye, it is hard to fathom the reality of impending doom.

Neil Frank, former director of the National Hurricane Center, has said that when he joined the staff of the organization in the 1970s, forecasters might expect a storm to veer off its projected course by as much as 120 nautical miles within a twenty-four hour period. By the time Frank retired twenty-five years later, and despite the deployment of technically advanced spotter planes, satellite imaging, and sophisticated computer modeling, the standard margin of error still stood at 110 miles.

In other words, a storm predicted to hit Homestead, on the tip of the mainland, might twenty-four hours later actually devastate Key West. Or, one thought to be passing safely between the Yucatán Peninsula and Key West might suddenly change course and batter Key Largo and Marathon instead.

Today such maneuvers can be followed, at least, and some measure of warning flashed ahead to the unfortunate new targets. In 1935, however, while weather experts had access to charts that detailed the workings of previous hurricanes, and they knew when storms were likely to appear, they were ill equipped to say just where a new one might come ashore.

Much of what was known about hurricanes in those days derived from the work of Reverend Benito Vines, a nineteenth-century Jesuit priest who was one of the first meteorologists to specialize in hurricane forecasting. Reverend Vines had established a meteorological observatory at Belén College in Havana, and while he had access to little in the way of instrumentation, his predictions had become so accurate that some in the community regarded him as having supernatural powers.

As to what to do about the menace, Reverend Vines was as perplexed as anyone else of his day, however. If nothing else, he advised appeals to higher authority, albeit in a systematic manner: "Priests in Puerto Rico should recite the prayer *Ad Repellandat Tempestates* during the months of August, September, but not October. In Cuba it should be recited in September and October, but not in August."

If prayer did not ward off the storm, other options were limited, certainly when one watches its approach from a spot on a tiny island, its landmass barely above sea level on a good day. Even today, experts presume that it would take a minimum of twenty-four hours to evacuate the Keys, and that is with a major highway running from Miami to Key West, two access bridges linking Key Largo to the mainland, and a carefully networked system of civil defense and sophisticated early-warning systems in place, none of which existed on that Labor Day weekend when Ernest Hemingway caught a glance at his evening paper's headlines.

On Sunday, September 1, as that 1935 storm approached Andros Island in the Bahamas, less than one hundred miles from the U.S. mainland, it was packing winds of seventy-five miles per hour, what would be termed a minimal, or Category 1, storm today. Bulletins issued by the Associated Press suggested that by the following morning the storm was likely to hit Havana, one hundred miles south of Key West, and pass on westward into the Gulf of Mexico.

Cuba's Carta Blanca weather observatory predicted that the storm was likely to pass through the channel separating Cuba from the Keys. The U.S. Weather Bureau in Jacksonville issued a storm warning at nine-thirty Sunday evening, locating a "tropical disturbance" about 250 miles east of Havana "mov-

ing slowly westward, attended by shifting gales and *probably*
[emphasis added] winds of hurricane force over small area
near center." The bulletin ended by advising "caution" for vessels in the Florida Straits.

All relatively benign, by hurricane standards. But less than
forty hours later, when the storm suddenly veered northward
toward the Keys, it had become a Category 5 monster, off the
charts in terms of wind speed—two hundred miles per hour
and more—and to this day the strongest ever to hit the United
States.

Hurricane forecasters are fond of using numbers to try to
convey the fearsome potential of the storms: they will say that
a hurricane releases as much energy in a single day as would
the detonation of four hundred twenty-megaton H-bombs, or
enough to provide the electrical needs of the United States for
six months.

One of the more interesting statistics has to do with wind
force. A "minimal" 75-mile-per-hour storm has the capability
of propelling a shard of two-by-four lumber through a four-
inch concrete block wall, but they will also tell you that when
wind speed doubles, wind *force* actually quadruples. So that a
Category 4, 150-mile-per-hour storm carries winds of four
times the force of the lowest-level hurricane. And while most
people would characterize a day with 20-mile-per-hour breezes
as "windy," a 200-mile-per-hour storm has winds of one hundred times that force.

The Galveston Hurricane of 1900, the deadliest in history,
with some eight thousand lives claimed, packed winds in the
150-mile-per-hour range; while Andrew, in 1992, the costliest
hurricane in history, with $25 billion in damages, was also
officially labeled a Category 4, 155-mile-per-hour storm.

Given what was coming at them on Labor Day of 1935, then, residents of the Keys could only hunker down and pray for the best.

※　◙　※

In the case of Hemingway and his fellow citizens of Key West, the approach seemed to be working. As Hemingway wrote:

> . . . *a little after two o'clock [the storm] backs into the west and by the law of circular storms you know the storm has passed over the Keys above us. Now the boat [Pilar] is sheltered by the sea wall and the breakwater and at five o'clock, the glass having been steady for an hour, you get back to the house. As you make your way in without a light you find a tree is down across the walk and a strange empty look in the front yard shows the big, old sapodillo tree is down too. You turn in.*
>
> *That's what happens when one misses you.*

※　※　※

Whether by divine intervention or some lesser force, Key West had escaped the worst of it. But for those thousand or so residents and workers caught in the Middle Keys—including rescue train engineer J. J. Haycraft, the seventy-four members of the extended Russell family, and the six hundred or more veterans working on the Overseas Highway—prayers seemed not to be getting through.

Certainly prayers were not helping seventeen-year-old Bernard Russell, who had been forced to flee the lime-packing shed on Upper Matecumbe, where he and several members of his family had hoped to find shelter from the raging storm.

They did not want to venture out into the raging storm, of course, but once the tides had risen to the level of that elevated building's floorboards, there was no choice—the walls could easily collapse and crush them all.

Once outside, Bernard Russell continued to shout to his twenty-one-year-old sister that, as the stronger, he should be the one to carry her infant child through the maelstrom. But he was arguing with a desperate mother not about to loosen her grip upon her child.

Within moments, Russell had felt a terrific gust of wind sweep over them. Instinctively he had thrown his arms about his sister and her baby, fighting to guide them away from the rising waters.

Scientists still debate whether true tornadoes are spawned within the bands of the most powerful hurricanes. Bernard Russell can tell you that he has no such doubts. For a whirling vortex had snatched him and his sister and her baby off the ground as if they were twigs, and was now spinning them about in an ever-expanding circle.

Russell struggled, but he was up against a force stronger than the fiercest human intent. He felt his hands loosen at his sister's shoulders and saw the look of panic in her face as she was pulled away, still clutching her baby. Though he strained to reach her, his arms could barely move against the force of the cyclonic winds.

As the vortex turned, she grew farther and farther away, until suddenly she and her baby were gone. In the next moment Russell found himself flying across the waterlogged packing-house grounds, just one more scrap of debris tossed by the indifferent storm.

By the time the winds released him and he could struggle to his feet, there was no trace of his sister or her child. "It was like looking in a bottle of ink," Russell was to tell *Miami Herald* reporter Nancy Klingener in a 1995 interview. "You could see nothing. The winds are howling. And the rains are pounding. It was chaos. We were raked with trees or big pieces of houses or whatever else was coming by."

As he reeled about the wind-blasted darkness, searching for his sister and other members of his family, Russell felt his foot plunge into a tangled deadfall. In the next moment the heavy mass shifted in the wind, pinning his ankle, and Russell realized he was trapped. The winds were so strong by this point that he had to turn his back and cup his face in his hands in order to breathe.

"It felt like eternity," he says. "It could have been thirty minutes. It could have been two hours. Time was nothing then."

At one point he sensed the waters receding beneath the tangle of brush where he was pinned, and thought that the worst was finally over. Then, suddenly, a chunk of siding torn from a building careened through the darkness and slammed atop the brush pile, driving Russell facedown into the water. He struggled, but the weight was too much. He was gulping seawater now, his arms floundering. It was over, then, he thought. All that was left was to die.

⊠   ⊠   ⊠

Today's Saffir-Simpson Scale is a chart developed by hurricane researchers that classifies storms by the level of wind strength and also summarizes the likely effects of each. According to Saffir-Simpson, those who experience a Category 5 storm, the strongest on the scale, can expect the following:

"Complete roof failures on many residences and buildings. Some complete building failures. Major damage to all structures located less than fifteen feet above sea level. . . . Intensive winds continue far into inland areas."

The threshold wind speed for such a storm is 155 miles per hour, and a tidal surge of eighteen feet can be expected to come ashore in advance of those winds. Only two storms of such intensity have struck the United States in the twentieth century. One was Camille, which came ashore near Biloxi, Mississippi, in 1969, with winds of approximately 175 miles per hour and a final death toll of more than 250. The other was the unnamed Labor Day storm of 1935, with its winds estimated at more than two hundred miles per hour (fewer than 3 percent of *tornadoes* ever generate winds of more than 206 miles per hour, according to experts).

The prospects for those caught on the Matecumbes, in what had become the storm's crosshairs, were even more grim when the nature of the "shores" being borne down upon were considered. There is no land in all the Keys that rises more than sixteen feet above sea level, and most of it lies far below that, at an average of three or four feet.

As for escape "far inland," that is a laughable concept in the narrow island chain. For Bernard Russell and his family, escape far inland had meant a dash through driving rains and wind a hundred yards or so to a lime-packing shed near the six-foot-high railroad bed.

And for the men working on the Overseas Highway, the outlook was just as dire. The "vets" as the highway workers had come to be called by the locals, were in truth largely World War I veterans who had marched upon Washington a few years before, demanding payment of bonus monies promised by Congress but never delivered. Although President Hoover had dispersed the "bonus army" encampments about the Capitol with armed troops and tear gas, Franklin Roosevelt had developed a more humane response, or at least that had been his intention when his relief agencies provided them employment in the Keys.

Unfortunately, however, these men were not being supervised by the Keys-seasoned and wary FEC project engineers, but by officials of the Federal Emergency Relief Administration, bureaucrats who had relatively little knowledge of what they were getting into. The three work camps that had been established (from Windley Key, MM 86, down to the end of Lower Matecumbe, MM 73) were not the sturdy, reinforced barracks of the FEC, but were comprised primarily of tents and flimsy temporary buildings that could easily be taken down and moved along as the crushed-coral roadbed progressed down the Keys. Certainly anyone working on the Keys would have a glimmering of what dangers a hurricane posed, but there was no Meredith or Krome in charge, and no longer any well-worked-out contingency plan in place.

Worse yet were those preliminary Weather Bureau reports downplaying the threat of the storm. Headlines in the Sunday edition of the *Miami Herald* made no mention of the impending storm: MIAMI TO OBSERVE LABOR DAY HOLIDAY was its blithe banner, with the subhead "Program Includes Parade, Sports, Picnic and Addresses." The forecast called for showers later that day, and "probably" for Monday as well. There was concern, then, in the highway work camps, but it was not great concern, for no one there could possibly have imagined what was about to happen.

※   ※   ※

Because it was a holiday weekend, and those who could afford it might well have been expected to have fled the work camps for the pleasures of Key West or Miami, some difference of opinion exists as to just how many of the 684 men assigned to the building of the highway were in those tents and shanties on that fateful Monday, but everyone agrees that there were hundreds still there. And now that the winds were rising, the rain had begun, and the whitecaps had begun to cover the island, the men were desperate for the arrival of that promised rescue train.

It was nearing 8:00 P.M. when Old 447 approached the Islamorada station on Upper Matecumbe Key (MM 82), but there had been no power to feed the building lights or approach signals for hours now. With waves sweeping over the seven-foot-high right-of-way and the wind-whipped sheets of rain intense, Haycraft was literally traveling blind.

Even once he spotted the group awaiting him, Haycraft was not quick to bring 447 to a stop. Understandably, the refugees had gathered in the lee of a cluster of buildings—the station house, the post office, and a warehouse—but Haycraft sensed that the winds might bring everything crashing down upon them at any moment.

Despite the frantic cries of the refugees who thought he was passing them by, Haycraft kept his hand on the throttle until 447 had passed more than a quarter of a mile beyond the station, at a point where the landmass was widest, and no buildings teetered overhead. "I think now that if I had stopped it at any other point, the train would now lie at the bottom of the ocean," Haycraft told a *Miami News* reporter.

There he waited while the pathetic figures chased down the track in his wake, dimly lit by the engine's headlamp, arms waving desperately. They'd been screaming with everything they had, imploring Haycraft to stop, but no human voice could overcome the engine's noise or the awful, unending roar of the storm.

"It's very difficult to imagine what it was like to be in two-hundred-mile-an-hour winds," says John Hope, Tropical Weather Expert for The Weather Channel. "Of course, it would just tear all substandard structures apart, they couldn't stand up, and a person couldn't stand up, either. I expect they had to crawl to try to get on the train because the wind was just enormous. Very difficult to breathe, a lot of sand blowing around in their faces, it was just unbelievable."

For five minutes after he had brought his engine to a halt, Haycraft watched men, women, and children struggling past his cab to the passenger cars, where crewmen pulled them frantically aboard. In that brief time, it might have seemed that there was hope, that all his efforts could lead to some small victory, but that was when he felt that terrible rumbling beneath his feet and stared out at the impossible tidal wall bearing down.

❧   ❧   ❧

Records indicate that the system was indeed a small one, with an eye no more than eight miles wide, and principal storm bands perhaps thirty miles across. But its compact size is also what gave the storm its legendary strength.

No wind-speed-measuring instruments survived, but subsequent engineering analysis of the damage left behind suggested that gusts inside that tightly packed circle ranged from 150 miles per hour at its onset to over 200 on the storm's trailing edge. Some experts have argued that gusts could have reached 250 miles per hour in some pockets of the storm, and they like to point out that the barometric readings that were taken at the fringes of the storm were of levels normally found inside the cones of tornadoes.

According to Keys historian Jerry Wilkinson, the legendary

barometric reading of 26.35, which lends further credence to the wind-speed estimates, was recorded not at a weather station but by Captain Ivan Olsen, who rode out the storm off the coast of the Matecumbes in his boat. Though the readings had fallen off the scale of Olsen's barometer, he used a knife to mark the brass faceplate as the mercury continued falling. Though at times the towering waves threatened to swamp his craft, causing Olsen to wonder if the markings would constitute a posthumous record of his hellish night, his battered craft held together, and scientists were able to calibrate the readings in the aftermath.

Most feel that the tidal surge that swept up out of the ocean that evening was eighteen to twenty feet high, though the September 1935 issue of the *Monthly Weather Review* suggests a more chilling possibility:

"The track and crossties of the railroad were in one stretch washed off a concrete viaduct thirty feet above ordinary water level. . . ." In any case, "reports agreed in the description of the great rapidity with which the rise of the sea came in from the southern side of the Keys as a 'wall of water'. . . ."

The winds were so strong that the waters covering the shallow reef were in fact being displaced, an ocean literally picked up and swept ashore—or what had been shore. The mighty wave fed upon itself, doubling and redoubling in size, until it had become a thundering, otherworldly mass that could dwarf an entire island, and swallow a train as if it were nothing at all.

Subsequent inquiries have suggested that the rescue train was intended for dispatch from Miami on Sunday, the day before the storm hit, and was to have remained on standby in Homestead, with a crew at the ready, in case it should be needed for a quick dash down to Marathon, no more than an hour away. But bureaucratic snafus in Washington prevented arrangements from being finalized with the railroad offices in Miami.

By the time the gravity of the situation was apparent and the frantic SOS went out on Monday at 2:35 P.M.—from a frantic work-camp supervisor in Islamorada—railroad crews had dispersed for the holiday, and the delays in assembling a train became inevitable.

Little of that mattered to J. J. Haycraft, of course, as the hellish winds howled and the terrible "wall of water" rushed toward him. What he was experiencing was beyond imagination, beyond anyone's power to prepare for.

He opened his throttle wide, in a desperate attempt to save the human cargo he had on board, but nature had other plans. The engine lurched forward only a few feet before it shuddered to a halt. The train's conductor, J. F. Gamble, flung himself into the cab, his uniform soaked and dripping, to report the worst: one of the hundred-ton boxcars at the rear of the train had been toppled by the wind and waves, automatically locking the air brakes on the entire chain. They were frozen in place, then, as the tidal surge advanced. And as far as Haycraft could envision, they were all as good as dead.

⌧   ⌧   ⌧

Survivors at the fringes of the storm told chilling tales of men disemboweled by jagged sheets of roofing torn loose by the wind, of skulls crushed by fifteen-pound boulders flying through the air like pebbles. The darkness was often interrupted by strange flashes of "ground lightning" a phenomenon generated by the wind lifting millions of sand granules into the air, where their clashing created eerie static charges. "It was as if millions of fireflies were swarming," one witness said.

Deadly fireflies, in this instance. Some victims who could not find shelter were later found where they had been pinned against piles of debris, their faces literally blasted down to bone by the driving sand.

Melton Jarrell, the workman from Camp 5, located near MM 79, who'd had his leg pinned by a thousand-pound section of rail and was ready to cut his own foot off to escape drowning, had passed out from pain and shock before he could carry out his gruesome plan. When he awakened, he realized that the tidal surge that had dropped the rail on his leg had now somehow managed to free him from his trap. It might have been a welcome realization, had he not found himself being swept out to sea.

He was tumbling through wind-whipped waves, his efforts to swim in the powerful currents nothing but a waste of energy. What kind of luck had he had? he was thinking. In the next moment he felt himself crash against something hard, then felt an impact at his head. Finally there was darkness once again.

Lloyd Fitchett, who'd been quartered in Camp 1, some seven miles north of Jarrell, on Windley Key, had run to the railroad embankment, hoping to escape the rising waters. "Buddies all around were shouting in panic," he told reporters. " 'Give me a hand, buddy. Save me. I'm drowning.' These were mingled with the groans from men already too far gone to cry aloud. I fought hard to keep my head out of water and inch by inch managed to creep to higher ground."

When he found a telephone pole, Fitchett climbed as high as he could above the rising waters. "I took my belt off and strapped myself to it. I heard a swishing noise, then a shriek, and realized I had been hit by flying debris. I learned later it was the roof of the barracks that had fallen on my chest.

A barrage of stones kept hitting me all over the body and then I partly lost consciousness. I hung on through the night in a semi-dazed condition and when daybreak came I could see the bodies of my dead comrades all around me. I counted fifteen."

Witnesses reported seeing a roof lift up whole from one house on Windley Key, flying off in the next instant as if it were a hat tossed by a careless giant. Moments later the walls of the house collapsed, disappearing before an onrushing tidal surge. Sofas, chairs, tables, and household goods of all sorts churned by in the raging current, followed by, of all things, a *piano*.

As if that sight were not bizarre enough, onlookers realized that there was a desperate woman draped over the piano's leading edge, her arms clutching as if she'd meant to keep her most prized possession from being carried away. As they gaped, woman and piano rushed by at incredible speed, hurtling two hundred yards inland before the furniture-bearing breaker finally crashed down. The wave broke against the railroad embankment, and the massive piano fell atop the body of its owner, crushing her.

J. E. Duane, the caretaker of the FEC's Long Key Fishing Camp, described his own fight for survival to reporters for the *Key West Citizen*:

By nine-thirty that evening, waters had risen to nearly twenty feet on the seaward side of the island, and the house where he and other employees had taken shelter seemed ready to collapse. During an unexpected lull, Duane and a coworker ventured out onto the porch of the house, where they were astonished to find the skies above them clear, the winds gentle, the stars shining brightly.

The calm seemed impossible, but soon enough, Duane realized that it was only the eye of the storm passing over them.

As he and his companion made a dash for a more solid-looking structure on higher ground sixty feet or so away, the winds resumed as abruptly as if a switch had been thrown, only this time with even greater force. Before they could make it to the other house, Duane and his companion found themselves swimming through waist-deep water.

Even as others in the house dragged them inside, the waters followed, flattening doors, pouring through shattered windows, rising to chest level within moments, and threatening to tear the entire structure off its foundations and out to sea.

Among those trapped in the building were four infant children. For more than half an hour, while the water rose, the men among the group held the children aloft to keep them from drowning, praying all the while for the storm to end.

⊠   ⊠   ⊠

Jay Barnes, author of *Florida's Hurricane History*, recounts this chilling, moment-by-moment account of the Weather Bureau's observer on Long Key:

"10:15 P.M.—The first blast from SSW, full force. House now breaking up. . . . I glanced at barometer which read 26.98 inches, dropped it in water, and was blown outside into sea; got hung up in broken fronds of coconut tree and hung on for dear life. I was then struck by some object and knocked unconscious.

"Sept. 3: 2:25 A.M.—I became conscious in tree and found I was lodged about 20 feet above ground. All water had disappeared from island; the cottage had been blown back on the island, from whence the sea receded and left it with all people safe."

That cottage was the very one where J. E. Duane had run for shelter, hours before, and where those four children had been held aloft by men and women who sobbed and cursed and prayed for deliverance. In one case at least, they had been successful.

The weatherman and the others inside the cottage were among the lucky ones, and so, as it turns out, was Melton Jarrell, who cheated death a second time that night. He was plucked unconscious from the top of a thirty-foot tree at the bay's edge, where the great tidal surge had left him.

The fates would not be quite so kind to J. J. Haycraft and his stalled rescue train, however. Engine 447 was a workhorse, built in Schenectady, New York, in the 1920s, designed for duty and not for grace. At just over 320,000 pounds, the engine gripped its rails as if the gravity of Jupiter were pressing upon it (imagine four Cadillac automobiles parked end to end, then piled twenty high; next, imagine that eighty-car stack crushed down to the height of a Greyhound bus, and you get some idea of the density). Still, as the water wall slammed down, Haycraft felt as if he were being tossed through frothing rapids in a birch-bark canoe.

There was a great lurch as the wave toppled the remainder of the eleven cars attached behind the engine. The linked cars went over sideways like toys, the windows of the passenger cars smashing inward under the tremendous force of the wave, the interiors filling almost instantaneously with water.

Inside, scores of men, women, and children who had thought themselves safe only moments before now found themselves trapped in water-filled coffins. In the surging darkness, there was no way to tell up, down, or sideways. Life had been reduced to the amount of time one might hold a single breath against the press of suffocating water.

Desperate parents groped madly for children torn from their grasp. Panicked men flailed blindly, their limbs tangling with those of others clawing just as wildly in return. As best we know, no one drowns with dignity.

A few did manage to escape the toppled cars through the shattered windows. Ironically, most of those "lucky" ones were swept out into the storm-tossed seas to die.

Engine 447, however, proved too heavy for even that monster wave to overturn. Miraculously, Haycraft and the crewmen in the engine's cab—among them conductor Gamble and fireman Will Walker—came up from the battering of the waves to discover that they were saved.

But for forty miles flanking that single, sixty-foot stretch of track upon which 447 still sat, the roadbed of the Key West Extension had been obliterated, as had everything else in the path of the storm: the station house where so many had fled to find shelter, solidly built Conch homes where others had cowered, even the trees that some had lashed themselves to in desperation. For forty miles, including the spot where the Long Key Fishing Camp had stood—and that where fellow visionary George Merrick had planted his last hopeful development, the Caribee Club—the earth had been scoured clean.

"The Florida East Coast Railway is a total wreck . . . tracks have been picked up and tossed aside, sometimes fifty yards from the roadbed," wrote one stunned reporter who had made his way to the site by boat.

"The trestles through the cuts are ruined. The hospital building at Camp No. 1 was swept so completely that not a splinter except the concrete base remains. Not a building stands there. . . . The foliage literally has vanished . . . everywhere one sees bedding, clothing and other bits of household necessities clinging to the brush, almost as if laid out to dry. Always it is high in the bushes, almost above a man's head. . . . The entire mass of lumber used in construction of all the homes and cottages along the east coast of the islands lies high in the underbrush fully 300 yards inland. You'll go a long way before you see such wreckage again."

From a distance, there seemed nothing left but tumbled rock, shattered stumps, and, here and there, a pile of rubble or a twisted rail. One had to look more closely to find the bodies.

⌧   ⌧   ⌧

Trying to paint a picture of hurricane devastation for those who have not seen such a landscape firsthand is a traditionally elusive task. In 1909, when Key West tobacco company executives were finally able to send telegraph accounts to their New York counterparts, describing the catastrophic effects the storm had had on their holdings, most executives dismissed the cables as "impossible canards."

Following Hurricane Andrew's assault on South Miami–Dade County in 1992, then-president George Bush flew over Homestead and its environs, and, even after that bird's-eye view of things, returned to Washington, still undecided whether federal relief funds were truly necessary. Outraged community leaders demanded that Bush return for a street-level assessment. Following a few hours of picking his way through the devastation that often required military half-tracks to traverse, a pale and chastened Bush vowed to set every mechanism of aid at the government's disposal into motion.

Following the 1935 storm, the first doctor to arrive in the Matecumbes was G. C. Franklin of Coconut Grove. He discovered the bodies of thirty-nine men in the first tangle of debris he encountered on shore, but even such statistics could not convey the feeling of despair that enveloped the islands in a palpable shroud.

"I saw a man, big powerful man, sitting upright on the road," wrote Jack Bell, a reporter covering the storm's aftermath for the *Miami News*. "He wore a blue denim jacket and overalls, nothing else. Beside him sat a little boy of about five, his head wrapped in a great ungainly bandage, a shirt ripped open.

"The man sat staring into space. 'What is your name?' I asked. He lifted his head and stared at me.

" 'Don't you remember your name?' I asked, trying to help him. Still he stared vacantly at me, saying nothing."

Ernest Hemingway, pinned in Key West by residual winds until the second morning after the storm, joined one of the first rescue parties to reach the Middle Keys and did his best to evoke the horrors he encountered. "When we reached Lower Matecumbe, there were bodies floating in the ferry slip," he writes in *New Masses*. "The brush was all brown as though autumn had come . . . but that was because the leaves had all been blown away. There was two feet of sand over the island where the sea had carried it and all the heavy bridge-building machines were on their sides. The island looked like the abandoned bed of a river where the sea had swept it."

Soon enough, he encountered even worse: "The railroad embankment was gone and the men who had cowered behind it . . . were all gone with it. You could find them face down and face up in the mangroves. . . . Then further on you found them high in the trees where the water had swept them . . . beginning to be too big for their blue jeans and jackets that they could never fill when they were on the bum and hungry."

There were anomalies, as there always are in a storm's aftermath, oddities that one wants to fasten on as evidence of a more benign universe, but rarely can. "On the other hand," Hemingway notes, "there are no buzzards. Absolutely no buz-

zards. How's that? Would you believe it? The wind killed all
the buzzards and all the big winged birds like pelicans too. You
can find them in the grass that's washed along the fill."

Another mind-boggling account is to be found in *Florida's
Hurricane History*: One victim was found the day after the
storm, impaled by a piece of two-by-four that had passed completely through him, just beneath his ribs and somehow missing his kidneys. The man was still living and appeared calm as
a doctor prepared to remove the length of timber. The doctor
offered the man a shot of morphine to dull the pain of the pro-
cedure, but the man refused, opining that the operation was
sure to kill him. That being the case, the man reasoned, he
would rather have two beers instead. He was given the beers,
which he drank, and then said to the doctor, "Now pull." The
doctor pulled the timber out. And the man died.

In a letter to his editor, Maxwell Perkins, Hemingway presents the most disturbing images of all, in terms that would
have made it impossible for any publication of the day to
print:

> *Max, you can't imagine it, two women, naked,
> tossed up into the trees by the water, swollen and
> stinking, their breasts as big as balloons, flies
> between their legs. Then, by figuring, you located
> where it is and recognize them as the two very nice
> girls who ran a sandwich place and filling-station
> three miles from the ferry. We located sixty-nine
> bodies where no one had been able to get in. Indian
> Key absolutely swept clean, not a blade of grass, and
> over the high center of it were scattered live conchs
> that came in with the sea, crawfish, and dead*

morays. The whole bottom of the sea blew over
it . . . we made five trips with provisions for sur-
vivors to different places and nothing but dead men
to eat the grub. . . .

The official Red Cross death toll was 408, but most agreed
that the official count was low, that the final tally would never
be known, owing to an uncertain Keys census and a general
laxity in FERA's record keeping. The Islamorada coroner put
the figure at 423, but many informed estimates quickly suggested the total was over 600, for many bodies of the missing
were never found, and there was no way to know just how
many of the vets were in the camp that day.

Whatever the final count, it is generally agreed that more
than half of the thousand or so residents and workers caught
on the Matecumbes that day lost their lives. (One measure of
the storm's ferocity comes from the dire observations of
present-day weather experts who try to attune coastal residents and visitors to the magnitude of hurricane threats: were
a storm as powerful as the Labor Day hurricane of 1935 to
strike Key West today, they say, the 22,000-person island
would likely be wiped as clean of life and property as the
Matecumbes were years ago. Were such a storm to strike a
major population center such as Miami, property damage
would likely outstrip Andrew one hundred times over.)

In any case, the full extent of the devastation will likely
never be known. Some twenty years later, an Islamorada developer digging fill out of a rock pit unearthed three automobiles
bearing out-of-state license plates dated 1935, the skeletons of
their occupants still resting inside. And to this day, those pok-

ing about one of the hundreds of tiny, uninhabited islands dotting Florida Bay will sometimes uncover remains suspected to be victims of the Labor Day storm.

⊠   ⊠   ⊠

Stunned by the magnitude of the losses, and eager to appear compassionate, Roosevelt's government issued initial orders that all of the bodies of the veterans would be removed to Arlington National Cemetery for interment. But it soon became clear that the great numbers of the dead made such a plan impractical. The bodies would be given proper burial in Miami, came the next decree.

One hundred sixteen bodies were taken to Miami, as it turned out. But even that undertaking would prove impossible to carry to conclusion, for the tropical heat and humidity were causing an unexpectedly rapid decomposition of the victims' flesh. State health officials, fearing the outbreak of an epidemic, ordered the bodies burned by the attending national guardsmen, and several clashes ensued when local residents tried to pull the bodies of relatives out of the funeral pyres— fed by burning tires and upturned railroad ties—for private ceremonies.

⊠   ⊠   ⊠

Here and there had been pockets of survivors, though, given the context of these lives, "survival" might put too positive a spin on the matter.

Thirteen of the vets who had taken leave of the camp for the Labor Day holiday were discovered AWOL in Key West, the Tuesday following the storm. They had traveled down to the city as part of a group of fifty, and had decided to stay on, despite the fact that they were due back at work on Monday.

Their dereliction would likely have cost them their jobs; but unlike the thirty-seven dutiful colleagues who had returned to camp on Monday, the thirteen AWOL still had their lives.

One newspaper account told of a ten-year-old girl found lying alongside the ruined highway in Islamorada, critically injured. The girl's clothing had been entirely ripped away by the winds, her body still bleeding from wounds she'd suffered from the unending barrage of debris that had darkened the skies a day before. As rescuers loaded her onto a launch for transfer to a hospital, she begged them to search for her family, who lived in a house nearby. No one had the heart to tell her that her mother and father had died, along with her four brothers and sisters, one of them seventeen months old.

Elsewhere, rescuers found three men wedged amid an outcropping of boulders, all of them apparently drowned. The workers were about to turn away, when one of the "dead" men stirred. As the man was revived, he recounted the tale of being flung into the rocks during the storm. A blow to his head had incapacitated him, and he had awakened to find his face only inches away from those of his two dead companions. He'd been staring back at their sightless eyes for nearly a day before his rescue.

And not far away, Bernard Russell had awakened atop that tangled deadfall where he'd been pinned, to realize he had not drowned after all. The waters had receded a bit and the winds had died back, enough for him to breathe, at least. In the dim early light, he was able to work his foot free and climb down from the uprooted mass of trees that had saved him.

Then he heard a voice nearby—it was someone who'd been injured, now calling out for help, he realized. Trying to orient himself in the still-powerful storm, Bernard Russell fought his way toward the sound. He'd gone only a few feet when he bumped into a man staggering in the opposite direction. The two backed off to stare at each other through the gale—and Bernard Russell felt his knees weaken in relief. Just as he was about to give up hope, he found himself staring at his father, alive.

John Russell had lost his shoes in the storm and had cut his feet terribly as he scrambled over wreckage in his own desperate search for his family. Bernard sat him down on the trunk of a fallen tree, then scavenged some sheeting and sponges from one of the wind-whipped piles of debris that lay everywhere. By now the skies were beginning to lighten, and the two could make out a railroad car tumbled on its side not far away. Bernard bound up his father's feet as best he could, and then helped him toward the fallen car.

The two men climbed through a window of the train, where they stayed until light came and the winds finally began to lessen. The moment it seemed practicable, Bernard pulled himself up and out, and began the search for others of their family.

The first body he stumbled over was that of his six-year-old cousin. His mother, Louise, had been drowned, as had the sister he had tried to protect, along with her child. Bernard's uncle Clifton had survived, but he had also lost his wife and four of his five children, including a daughter whose body was discovered forty miles away, on a scrub island near Cape Sable, where the railroad had once planned to launch itself off the coast. Still clutched in the young woman's arms was the body of her dead baby.

"No way she could have held on to that baby if they'd been in the water all that distance," Bernard Russell says. "They'd been lifted up and blown there by the wind."

In all, sixty-three members of the Russell family died that night. Only eleven lived to remember its horrors.

To this day, Bernard Russell recalls wandering through the hellish aftermath, wondering if he was, in fact, one of the lucky ones. "Everything was flattened," he says, "everything was gone. It was a different world. I went back to where our house had been and there was nothing. Just an empty space.

"A numbness comes over you," he continues, repeating the familiar refrain of the hurricane survivor, "seeing one terrible thing after another. It's hard to explain if you haven't experienced it."

One memory remains clear in Russell's mind, though. As he and his father stood together surveying the devastation, Bernard glanced up at his father in despair. "Dad, what are we going to do?" he asked.

"All we have left is this property," John Russell replied, waving his hand about the ruined landscape surrounding them. "What do *you* want to do?"

Bernard Russell looked about, trying to comprehend what had happened, trying to reason what possibilities might remain. When he finally answered, it was in a way that Henry Flagler would surely have approved. "I guess we'll have to start over," he said. "We're going to dig in and start over."

Dig in, the Russells did. The Red Cross built for them—as they did for twenty-seven other survivors—a fortress of a home with eighteen-inch poured-concrete walls on a five-foot-high raised foundation, not far inland from the Atlantic.

Bernard's father, who had become Islamorada's first postmaster after the train arrived, stayed on, and after a tour with the Coast Guard and the Army, Bernard would return as well. He and his wife, Laurette, herself a survivor of the storm, still live in the house the Red Cross built, married sixty-six years at the time of this writing. For years Russell worked as a carpenter, and in his off-time he served as a civil defense official and founded the island's first fire department.

He acknowledges that the storm changed him: "I suppose it made me more aware of the terrible things that can happen. When I came back after the service, I figured I ought to try and help people realize.

"My dad used to say that the railroad changed Islamorada from an island to a part of the world," Russell says, but he harbors no resentment that the rescue train couldn't reach his family on the day that it mattered most. "That was no one's fault," he insists. "That's the thing about hurricanes. No one could have imagined what was going to happen."

◻   ◼   ◼

Bernard Russell's capacity for forgiveness seems remarkable, though the passage of time may well have helped in that regard. Certainly others felt less charitable, especially in the storm's immediate aftermath.

Because several of the smaller bridges above and below Islamorada had been washed out by the hurricane, there was no access to the disaster area but by boat. Phone and telegraph lines had been washed away as well. For at least two days, most of the world was unaware of the magnitude of the disaster.

But when tales of as many as one thousand deaths began to circulate, and photographs of the obliterated Keys found their way into newspapers across the country, outrage grew.

"Who sent nearly a thousand war veterans, many of them husky, hard-working and simply out of luck, but many of them close to the border of pathological cases, to live in frame shacks on the Florida Keys in hurricane months?" Ernest

Hemingway demanded angrily, in his *New Masses* dispatch. "Who advised against sending the train from Miami to evacuate the veterans until four-thirty o'clock on Monday so that it was blown off the tracks before it ever reached the lower camps?"

From St. Augustine, Scott Lofton, who had been appointed as receiver for the financially troubled FEC, issued an emphatic defense of the railroad: the veterans stranded on the Keys could not have been evacuated successfully unless the rescue train had left Miami by 10:00 A.M. on September 2, he insisted. It was 2:00 P.M. that day before F. B. Ghent, director of the Veterans' Work Program, had placed a call to the FEC, and the fact that a train and crew had been assembled and dispatched from Miami within two and one-half hours was remarkable, in Lofton's eyes.

"The officials of the railway have co-operated with the FERA officials in every way possible since the veterans have been in camps on the keys, and in this instance exerted themselves to the utmost to get this special train out from Miami at the earliest possible time notwithstanding the many handicaps that existed," Lofton said.

Despite such statements, congressmen and senators had soon joined the chorus demanding answers to these questions. A congressional investigation of the Federal Emergency Relief Administration and the Weather Bureau was conducted, due in large part to Hemingway's widely publicized outrage, an undertaking that would require more than two years to conclude.

When it was over, the investigators had come to the same conclusion as Bernard Russell, however: "The evidence clearly shows that the tidal wave was entirely unexpected and that it was impossible to even anticipate the hurricane within sufficient time to ensure safety for those concerned."

It was not much by way of recompense for all those lives lost, the lives of men who worked for one dollar a day plus room and board to build a highway down the Keys. Little more was to come.

A memorial was built by the Works Progress Administration in memory of the victims, atop a crypt containing the remains and ashes of about three hundred. The structure, in Islamorada at MM 81.5, was unveiled before a crowd of five thousand on November 14, 1937, by nine-year-old hurricane survivor Fay Marie Parker. "Dedicated to the memory of the civilians and war veterans whose lives were lost in the hurricane of September second, 1935," reads the inscription. One could easily walk from Bernard Russell's house to read it.

⌗   ⌗   ⌗

Some hoped for yet another rebirth of the Key West Extension, pointing to the fact that despite the devastation of much of the low-lying roadbed, Flagler's mighty bridges had weathered the storm virtually unscathed. But already bankrupt and in the hands of a receiver, the FEC was in no position to rebuild anything, much less a project that *Scientific American* had once described as "one of the most difficult works of railroad construction ever attempted."

Desperate for any ready source of cash, the right-of-way was sold to the state shortly after the disaster for $640,000, a sorry return indeed on a project that had required nearly $30 million, seven years, and the labor—and in quite a few cases, the lives—of a forty-thousand-man workforce, as well as the determination of a visionary, to build.

24

A Fine, Improper Place

As early as 1928, Congress had passed legislation making provision for an Overseas Highway across the Keys, and by 1935 one could actually drive much of the distance between Miami and Key West, though an intrepid motorist would have to make use of several ferry services on his way to the Southernmost City. The vets who had died in the Labor Day storm actually had been involved in the construction of a section of highway linking Grassy Key and Lower Matecumbe so that one of those ferry links could be eliminated.

With the railroad blown away by the hurricane, and the FEC through with the Keys forever, the state of Florida stepped into the breach, determining to use the abandoned right-of-way and the still-standing bridge spans as the route for a highway through the Keys. The road was finally completed in 1938, though it took the Second World War and the reactivation of the Navy base to resurrect the moribund economy of Key West.

If Flagler and Jefferson Browne had been wrong in predicting that "the products of the West Indies and Caribbean sea will be ferried across from Cuba and taken by the railroad for distribution to all parts of the U.S.," and that "with the completion of the Nicaraguan Canal, Key West would be a port of call for no small part of the shipping of the world," they had been prescient in one way at least: "Sooner or later the thousands of tourists who are restlessly seeking a milder and more equable winter climate than the mainland affords will find in Key West their ideal."

Following the war, the Southernmost City once again became a thriving tourist destination, and though now the pirates and the wreckers are gone, as are the turtle-raising corrals and the sponge divers and the cigar factories and the Cuban rollers and their lectors, who read newspapers and novels aloud while they worked, and the blustery big-game-hunting author who loved it all—though all this has vanished, and though the shrimpers and the fishermen and the Navy boys are nearly gone as well, obdurate Key West remains.

Irreverent and, to some, irrelevant, the oddball city at the end of the line is still the "Last Resort." The sun still shines, the Gulf Stream still flows, and palms still wave in the island's balmy breezes. Tourists and sports fishermen still descend, the colonies of writers and artists endure, and Hemingway's house on Whitehead has become a museum where the curious and the devoted come by the hundreds and the thousands each day. T-shirt shops and rowdy bars thrive, cheek by jowl with the swankiest of resort hotels.

And in its quirkiness and beauty may be found the proper legacy of Henry Morrison Flagler. Though it is hard to imagine Flagler and his crowd brawling down a crowded Key West street during one of the seemingly endless festivals concocted by the city's tourism mavens, he was not averse to a good time, and most men of his ilk and day would never have done *most* of the things he did.

Had he been content to reinvest his Standard Oil dividends instead of spending them to invent Florida, Henry Morrison Flagler might be as synonymous with wealth and power as his former partner and sales associate John D. Rockefeller. But Rockefeller did the safe and sane thing, and Flagler built his Speedway to Sunshine.

As one writer of his time once put it, Henry Flagler went down to Florida "with its palms and red poinsettias, its white beaches and blue water, and so to speak, began life all over again." A joke of the day had it that the abbreviation of Fla. on any sender's letter was shorthand for Flagler.

He created a string of world-famous resorts all the way from St. Augustine to Key West, with a railroad to bring others to them, and when he had to do the impossible to reach that final destination, he rolled up his sleeves and he made it happen. Significantly enough, he did not ask Joseph Parrott if the Extension to Key West could turn a profit, but if it could be *built.*

Admittedly, he lived in a day before personal income taxes, and—despite his carping about Teddy Roosevelt—before anything vaguely resembling the scrutiny, environmental and otherwise, that today's movers and shakers must undergo. But as biographer Edward N. Akin asserts, Flagler, for all his stoicism, was not so much a businessman as a visionary in businessman's clothing. Or at least that is what he became during the second phase of his life.

Flagler, who had never traveled to Europe, who had never

been so far as California, found himself at age fifty-five, some-how arrived in Florida, in St. Augustine, and the result was transforming.

"It was the oldest city in the United States," wrote Edwin Lefevre in *Everybody's*. "He saw the old slave market, he saw the old *Spanish* fort; he saw the old city gates! He saw what you and I saw when we went to Pompeii or first gazed on the Pyramids! He saw palms—*palms!*—this man who had grown up in Ohio amid the wheat."

For Flagler, Lefevre said, "St. Augustine was a magic pool," and there the great man "steeped his soul in the glamour and romance of antiquity."

Whether Flagler came to Florida with a vision, or acquired it in the way that Lefevre suggests, he distinguished himself from his ilk, for the trouble with most visionaries is their lack of ability to make a vision whole.

By the time he arrived at the end of the continent, however, Flagler had plenty of experience with the practical. Not only could he dream a railroad across the ocean, he had the where-withal and the know-how to turn such a dream into concrete-and-steel-reinforced fact.

There are no more men like Henry Flagler, and there are no more dreams like his. Today we have software titans, and their minions who seek to bridge gaps measured in millimicrons and nanoseconds. Such accomplishment may be dizzying in its own right . . . but that kind of bridge-building pales in comparison to those that Flagler built across the Florida Keys.

Shelley's poem "Ozymandias" tells the fanciful tale of a wide-ranging traveler who encounters the shattered statue of a long-dead emperor tumbled to a barren desert landscape. The inscription on the blasted statue's base commands the traveler, "Look on my Works, ye Mighty, and despair!" though all there is to gaze upon are "lone and level sands." The enjoyment of the poem depends upon Shelley's implication that such grand dreams of the high and mighty are doomed to eventual ruin.

And so might a modern traveler find himself, halfway across the new Bahia Honda highway bridge, glancing off to eastward at the Stonehenge-like remains of that ancient railroad span still defying the Atlantic, perhaps thinking of Henry Flagler and, perhaps, tempted to buy into Shelley's argument.

But there is no stopping, not halfway across a busy highway bridge. And by the time the traveler reaches Key West, order is restored.

Acknowledgments

I AM INDEBTED TO many, of course, though to none more than John Blades, director of the Henry Flagler Museum in Palm Beach, as well as to the chief curator of that institution, Sandra Barghini, and to its archivist, Lisa Diamond. The holdings of the Flagler Museum contain virtually every extant document bearing on the building of the Key West Extension, as well as copies of the research, popular writing, and ephemera that have been produced in the years since.

I am also greatly indebted to the work of three Flagler biographers:

Sidney Walter Martin, whose *Florida's Flagler* was originally published by the University of Georgia in 1949 and reprinted as *Henry Flagler: Visionary of the Gilded Age* in 1998 by Tailored Tours Publications.

David Leon Chandler, whose *Henry Flagler* was published by Macmillan in 1986.

And Edward N. Akin, whose *Flagler: Rockefeller Partner and Florida Baron* was published by Kent State University Press in 1988.

Each deftly places a summary of the work on the Extension within the greater context of Flagler's life, and those interested in the whole will take great pleasure in those volumes.

For an overview of FEC history (as well for his cautionary advice on undertaking this project to begin with), I am indebted to Seth Bramson and his exhaustive compilation, *Speedway to Sunshine: The Story of the Florida East Coast Railway,* published by Boston Mills Press in 1984. I would also like to thank Jerry Wilkinson, president of the Historical Society of the Upper Keys, and Tom Hambright, Monroe County historian, for all their help in ferreting out valuable materials pertaining to this story.

For an overview of Florida Keys history, topography, customs, and quirks, no one has been more helpful than Joy Williams, whose so-called guidebook, *The Florida Keys: From Key Largo to Key West,* goes well beyond the definition. If Joy Williams's book is a "guide," then Julia Child is into "food prep."

Finally, thanks should go to Pat Parks, whose forty-four-page booklet, *The Railroad That Died at Sea* (Brattleboro, Vt.: Stephen Green Press, 1968; reprint Key West: Langley Press, 1996) so aptly and concisely summarizes Flagler's undertaking, the first to do so for the modern reader.

There are many other writers whose articles and books and stories have shed light on various aspects of the material herein. Because this book does not present itself as a work of traditional historical scholarship, however, and in order to lessen the burden on the general reader, I have not employed footnotes and have attempted to make appropriate reference to these writers and works in the context of this story. I have also appended a bibliography that details these sources and that may be of use to those interested in pursuing further research.

Selected Bibliography

In ADDITION TO THE exhaustive holdings of the Flagler Library and to those works already cited, the following are sources that have provided information for the writing of *Last Train to Paradise*. Where information unique to any single publication is referred to, a citation has been made in the text.

Books and Monographs

Barnes, Jay. *Florida's Hurricane History.* Chapel Hill, N.C., and London: University of North Carolina Press, 1998.

Bethel, Rod. *First Overseas Highway to Key West, Florida.* 1989.

Brian, Denis. *True Gen: An Intimate Portrait of Hemingway.* New York: Grove Press, 1988.

Browne, Jefferson B. *Key West: The Old and the New.* St. Augustine: The Record Co., 1912.

Chapin, George M. *Official Souvenir: Key West Extension of the Florida East Coast Overseas Railroad Extension.* St. Augustine: The Record Co., 1912.

Cox, Christopher. *A Key West Companion.* New York: St. Martin's Press, 1983.

Ellis, William E. *Robert Worth Bingham and the Southern Mystique.* Kent, Ohio: Kent State University Press, 1997.

Gallagher, Dan. *Pigeon Key and the Seven Mile Bridge, 1908–1912.* Marathon, Fla.: Pigeon Key Foundation, 1995.

Hersey, John. *Key West Stories.* New York: Knopf, 1994.

McCullough, David. *The Path Between the Seas: The Creation of the Panama Canal, 1870–1914.* New York: Simon & Schuster, 1977.

McIver, Stuart. *Hemingway's Key West.* Sarasota: Pineapple Press, 1993.

McLendon, James. *Papa: Hemingway in Key West.* Key West: Langley Press, 1990.

Murphy, George, ed. *The Key West Reader.* Key West: Tortugas Ltd., 1989.

Parks, Arva Moore. *Miami: The Magic City.* Miami: Centennial Press, 1991.

Pyfrom, Priscilla Coe. *The Bridges Stand Tall.* Marathon, Fla.: Pigeon Key Foundation, 1998.

Schulberg, Budd. *Sparring with Hemingway.* Chicago: Dee, 1995.

Tinkham, Todd. "The Construction of the Key West Extension of the Florida East Coast Railway, 1905–1915." Unpublished mss., Kalamazoo College Library, 1968.

Westfall, L. Glenn. *Key West: Cigar City USA.* Key West: Historic Key West Preservation Board, 1984.

Windhorn, Stan, and Wright Langley. *Yesterday's Florida Keys.* Key West: Langley Press, 1975.

Significant Contemporary Articles

Browne, Jefferson B. "Across the Gulf by Rail." *National Geographic* (June 1896).

Corliss, Carlton J. "Building the Overseas Railway to Key West." Talk before the Historical Association of Southern Florida, April 7, 1953. In *Tequesta* 13 (1953): 3–21.

Lefevre, Edwin. "Flagler and Florida." *Everybody's* (February 1910): 168–186.

Paine, Ralph D. "Over the Florida Keys by Rail." *Everybody's* (February 1908): 147–156.

Rockwell, John Maurer. "Opening the Oversea Railway to Key West." *Colliers* 48 (20 January 1912).

Venable, William Mayo. "The Long Key Viaduct." *Engineering Record* 56, no. 21 (23 November 1907): 558–560.

Significant News Articles

Klingener, Nancy. "The Dangerous Summer." *Tropic Magazine of the Miami Herald,* 3 September 1995, 7–12.

Klinkenberg, Jeff. "Killer in the Keys." *St. Petersburg Times,* 14 August 1991.

Sanders, William. "Survivor Recalls the Storm of 1906." *Miami Herald,* 24 April 1960, 5F.

Index

About the Author

Les Standiford is the author of eight critically acclaimed novels as well as several works of nonfiction. He has received the Frank O'Connor Award for Short Fiction and fellowships from the National Endowment for the Arts, the National Endowment for the Humanities, and the Florida Division of Cultural Affairs. He is a professor of English and director of the Creative Writing Program at Florida International University in Miami, where he has lived since 1981 with his wife and three children. You can visit his website and e-mail him at www.les-standiford.com.